MOLECULAR BIOLOGY OF ADENOVIRUSES

BY

L. PHILIPSON, U. PETTERSSON,
AND U. LINDBERG

1975

SPRINGER-VERLAG

NEW YORK WIEN

This work is subject to copyright.
All rights are reserved,
whether the whole or part of the material is concerned,
specifically those of translation, reprinting, re-use of illustrations,
broadcasting, reproduction by photocopying machine
or similar means, and storage in data banks.
© 1975 by Springer-Verlag/Wien
Printed in Austria by R. Spies & Co., A-1050 Wien

Library of Congress Cataloging in Publication Data. Philipson, Lennart, 1929–. Molecular biology of adenoviruses. (Virology monographs; 14) Bibliography: p. 1. Adenoviruses. 2. Molecular biology. I. Pettersson, Ulf, 1942–, joint author. II. Lindberg, Uno, 1939–, joint author. III. Title. IV. Series. [DNLM: 1. Adenovirus. 2. Adenovirus infections. 3. Molecular biology. W1 VI83 v. 14 / QW160.A2 P555m] QR360.V52. no. 14 [QR396]. 576'.64'08s [616.01'94]. 75-6658

ISBN 0-387-81284-9 Springer-Verlag New York-Wien
ISBN 3-211-81284-9 Springer-Verlag Wien-New York

Molecular Biology of Adenoviruses

By

L. Philipson, U. Pettersson, and U. Lindberg

Department of Microbiology, The Wallenberg Laboratory
Uppsala University, Uppsala, Sweden

With 20 Figures

Table of Contents

I. Introduction	3
II. The Architecture of the Virion	5
III. The Composition of the Virion	7
IV. The Adenovirus Genome	11
A. Physical-Chemical Properties	11
B. Homology between DNA from Different Serotypes	13
C. Infectivity of Adenovirus DNA	16
V. The Productive Infection	16
A. Early Interaction between Virus and Cell	17
B. Early RNA Synthesis	19
C. Late RNA Synthesis	21
D. Mapping of Early and Late Viral RNA on Adenovirus DNA	22
E. A Comparison between Cellular and Viral mRNA Production	26
F. Adenovirus DNA Replication	29
1. Characteristics of Adenovirus DNA during *in vivo* Replication	29
2. Replication of Adenovirus DNA in Isolated Nuclei	31
G. Translation	33
1. Early Proteins	33
2. Late Proteins	34
3. Modification of Viral Polypeptides	36
4. Translation of Adenovirus mRNA *in vitro*	36
H. Host Cell Macromolecular Synthesis	38
I. Control Mechanisms	40
1. RNA Synthesis and Processing	40
2. Translation	40
J. Assembly of Adenoviruses	42
1. Mechanism of Assembly	42
2. Defective Assembly	44
a) Arginine Starvation	44
b) Elevated Temperature	44
VI. Abortive Infections	45
A. Adenovirus Infection in Hamster Cells	45
B. Adenovirus Infection in Monkey Cells	46

VII. Adeno-SV 40 Hybrid Viruses	47
VIII. Adenovirus Genetics	51
IX. Cell Transformation	53
A. Cell Transformation by Different Adenoviruses	54
B. Viral DNA in Transformed Cells	54
C. Synthesis of Viral RNA	55
D. Properties of Adenovirus Transformed Cells	56
X. Tumour Induction by Adenoviruses	58
A. Induction of Tumors in Rodents	58
B. Do Adenoviruses Play a Role in Human Cancer?	60
XI. Biochemistry and Immunology of Adenovirus Structural Proteins	61
A. The Hexon	63
1. Morphology	64
2. Physical-Chemical Properties	64
3. Crystallization	66
4. Immunological Properties	67
B. The Fiber	68
1. Morphology	68
2. Physical-Chemical Properties	68
3. Immunological Properties	69
C. The Penton	70
1. Morphology	70
2. Physical-Chemical Properties	71
3. Immunological Properties	71
D. The Major Core Protein	72
E. Low Molecular Weight Proteins of the Virion	73
XII. Physiological Effects of the Structural Proteins	73
A. Hemagglutination	73
B. Neutralization of Adenoviruses	76
C. Possible Functions of Individual Adenovirus Proteins	77
1. Hexon	77
2. Penton	77
3. Fiber	78
4. Virus-Induced Non-Structural Proteins	79
XIII. Classification and Nomenclature of Adenoviruses	80
A. Adenoviruses from Different Species	80
B. Subgroup Classification	82
XIV. Adenovirus Infection in Humans and Animals	84
A. Human Adenovirus Infections	84
1. Pathogenicity	84
2. Epidemiology and Control of Adenovirus Infections	85
B. Infections in Animals	86
XV. Aspects on Adenoviruses as a Tool in Cell Biology	87
Acknowledgements	89
References	89

I. Introduction

In his biography "Arrow in the Blue" the author Arthur Koestler suggests ironically that the fate of an individual may be predicted by examining the content of the newspapers at birth. Adenoviruses were discovered in 1953 (ROWE et al., 1953; HILLEMAN and WERNER, 1954). At this time the Salk poliomyelitis vaccine was developed (SALK et al., 1954) and in the same year the discovery of the double helical structure of DNA (WATSON and CRICK, 1953) and the plaque assay for one animal virus (DULBECCO and VOGT, 1953) was announced. Thus, this new group of viruses was born with great hopes for progress in molecular biology and for the control of animal virus infections. In the short interval between 1953 and 1956 the adenoviruses were discovered, methods for laboratory diagnosis and serotyping were established, the epidemiology was clarified and a highly effective vaccine was developed and approved (for a review see HILLEMAN, 1966). Succeeding years showed, however, that the vaccines were contaminated with the oncogenic SV40 virus and that the adenoviruses themselves were tumorigenic.

Since the discovery of adenoviruses animal virology was developed into a quantitative science offering explanation for viral functions at the molecular level. Precise biochemical tools to characterize the genome and its transcription products as well as the structural proteins of these viruses are now available. Many of the pathways involved in control of the host cell and the viral genome during lytic infection with adenoviruses as well as the properties of the structural and the non-structural polypeptides synthesized during early and late productive infection are currently investigated. The dissection of the virion itself and its component parts has been rewarding since most of the structural proteins are soluble under non-denaturing conditions and available in sufficient quantities for structural, immunological and functional studies.

The name adenovirus was coined in 1956 (ENDERS et al., 1956) to designate this group of viruses isolated from the respiratory tracts of man and other animals. The adenoviruses are non-enveloped icosahedral viruses with genomes which are larger than those of papovaviruses, but less overwhelming than those of the T-even bacteriophages, the pox and the herpes viruses. The adenovirus chromosome has a molecular weight of about 23×10^6 and could thus code for 25—50 average sized polypeptides. Today more than 80 different adenovirus serotypes have been isolated from a variety of animal species (WILNER, 1969; WADELL, 1970) and all except the avian and some bovine adenoviruses seem to share one antigenic determinant. It was recognized early that the different adenovirus serotypes possess a high degree of individuality with regard to a number of attributes such as cytopathology, host range, hemagglutination properties, neutralization kinetics and oncogenicity. The human adenoviruses have been divided into subgroups on the basis of their ability to agglutinate rhesus monkey and rat erythrocytes (ROSEN, 1960) and on the basis of their oncogenicity (HUEBNER, 1967). Each subgroup contains several serotypes characterized by type specific antigens present in their capsids, as revealed by hemagglutination inhibition or neutralization tests.

In humans, adenoviruses cause primarily mild respiratory diseases, but conjunctivitis, myocarditis, enteritis, and lymph node involvement have also been reported (SOHIER et al., 1965). The epidemic character of the respiratory illness caused by some serotypes among military recruits has emphasized the need for the production of multivalent vaccines (HILLEMAN, 1966).

One biological aspect of the adenoviruses, which has received much attention during the last ten years, is their oncogenicity. In 1962 TRENTIN et al. (1962), discovered that human adenovirus type 12 (ad 12) induces tumors in newborn hamsters and HUEBNER et al. (1962), reported that ad 18 also has this property. Subsequently, a series of reports have confirmed the oncogenicity of human ad 12 and also described the oncogenicity of several other human and non-human adenoviruses for hamsters and other rodents (HUEBNER et al., 1963; RABSON et al., 1964a; YABE et al., 1964; PEREIRA et al., 1965; HUEBNER et al., 1965; GIRARDI et al., 1964; HULL et al., 1965; DARBYSHIRE, 1966; SARMA et al., 1965). The tumors usually have the characteristics of undifferentiated sarcomas, although malignant lymphomas have occasionally been observed (LARSON et al., 1965).

Transformation of *in vitro* cultured cells by an adenovirus was first demonstrated by MCBRIDE and WIENER (1964), who showed that the oncogenic human ad 12 could transform newborn hamster kidney cells. Subsequently, the transformation of rat embryo fibroblasts by the same adenovirus type was reported by FREEMAN et al., (1967a) Since then, several adenovirus types have been shown to cause *in vitro* transformation of rodent cells (FREEMAN et al., 1967b and c; VAN DER NOORDA, 1968a and b; MCALLISTER et al., 1969a and b; RIGGS and LENETTE, 1967; CASTO, 1969), and the tumorigenic properties of the *in vitro* transformed cells have been established. The biochemical studies of cells transformed by oncogenic viruses *in vitro* have progressed rapidly during recent years and much of our basic knowledge concerning growth regulation and tumorigenesis has been derived from studies on such *in vitro* systems (for a review see PONTÉN, 1971).

In addition to providing an insight about the changes involved in transformation of normally growing cells to tumor cells, the adenoviruses have become important in providing a convenient model system for studies of regulation of gene expression in eukaryotic cells. In the transformed cells, viral genes are apparently integrated in the host cell genome (GREEN, 1970), and results of studies on the expression of these viral genes would thus reflect the mechanisms used by the host cell in expressing its own genes. Recent developments in the analysis of adenoviruses reproduction indicate that *productively infected cells* also could serve as a convenient model system for studies on the synthesis of macromolecules in the eukaryotic cell. Several observations suggest that also under these conditions the adenoviruses mimic the host cell in its gene expression. After the infection the virus is rapidly established in the nucleus of the cell, where its DNA is transcribed and replicated (GREEN et al., 1970). The viral DNA is transcribed into large RNA molecules (GREEN et al., 1970; WALL et al., 1972; MCGUIRE et al., 1972), which by posttranscriptional events are cleaved into smaller mRNA molecules found associated with polyribosomes (GREEN et al., 1970; PHILIPSON et al., 1971; LINDBERG et al., 1972). Adenovirus mRNA is polyadenylated at the 3'terminus in the same way as cell mRNA. Messenger ribonucleoprotein particles (mRNP) isolated from polysomes during virus infection contain five major labeled polypeptides, four of which are identical in size to those found in the mRNP from uninfected cells (LINDBERG and SUNDQUIST, 1974).

Conditional lethal mutants of adenoviruses have recently been isolated. A large number of temperature-sensitive mutants of serotype 5 and 12 and CELO virus have already been described (WILLIAMS et al., 1971; LUNDHOLM and DOERFLER, 1971; ENSINGER and GINSBERG, 1971; SHIROKI et al., 1972; ISHIBASHI, 1970, 1971). More than 50 ad5 ts-mutants distributed between at least 17 complementation groups have been isolated and are currently undergoing biochemical and genetic analysis (WILLIAMS and USTACELEBI, 1971; RUSSELL et al., 1972a; WILKIE et al., 1973; WILLIAMS, personal communication). A genetic map of the virus chromosome based on two-factor crosses has already appeared (WILLIAMS et al., 1974).

During recent years new physical-chemical techniques have been developed which allow mapping of loci for specific functions on viral chromosomes. Thus, studies of adenovirus DNA utilizing liquid phase hybridization, restriction enzyme fragments and electron microscopic mapping are in progress. Adenovirus DNA of several types has been cleaved by restriction endonucleases into unique sets of fragments, which can be separated (PETTERSSON et al., 1973; MULDER et al., 1974). The DNA strands of at least three adenovirus types have been separated (LANDGRAF-LEURS and GREEN, 1971; PATCH et al., 1972), and in the case of ad 2 strand separation can be achieved also with isolated DNA fragments generated by restriction endonucleases (TIBBETTS and PETTERSSON, 1974; SHARP et al., 1974a). Hopefully, with the aid of these techniques we will soon learn how the expression of the viral genome is controlled both in productive infection and also in cells transformed by adenoviruses.

This rapidly expanding field has been reviewed frequently during recent years with emphasis on different aspects of adenovirus research. Relevant reviews of a general character are those of GREEN (1966), SCHLESINGER (1969), GINSBERG (1969), GREEN (1970), TOOZE (1973) and PHILIPSON and LINDBERG (1974). The characteristics of the structural proteins (PHILIPSON and PETTERSSON, 1973) and the biological properties of the capsid components (NORRBY, 1968, 1971) have also been reviewed. Reviews on the oncogenic and transforming capacity of adenoviruses have also appeared (GREEN, 1970; HOMBURGER, 1973; TOOZE, 1973).

II. The Architecture of the Virion

The adenoviruses are nonenveloped viruses with a diameter of 65—80 nm. The capsid is composed of 252 capsomeres arranged into an icosahedron with 20 triangular facets and 12 vertices as originally shown in the classical electron micrographs of HORNE et al. (1959). Figure 1 A shows a schematic drawing of the virion with the major components of the capsid indicated. Figures 1 B and C show electron micrographs which reveal the vertex region and the triangular facets of the virion, respectively. 240 out of the 252 capsomeres have six neighbours and are called *hexons* (GINSBERG et al., 1966) whereas the 12 capsomeres at the vertices

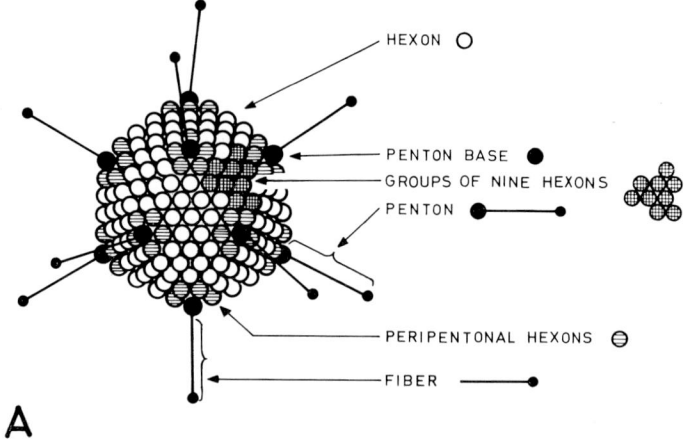

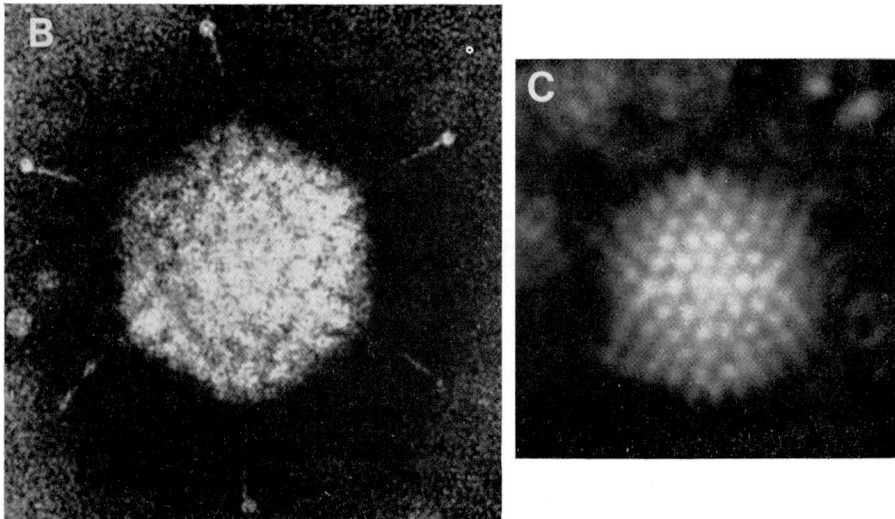

Fig. 1. Structure of the adenovirus capsid.
A. Schematic drawing showing the icosahedral outline of the adenovirus capsid and the location of various components. Reprinted from PHILIPSON and PETTERSSON (1973)
B. Electronmicrograph showing the ad 5 virion contrasted with sodium silicotungstate (VALENTINE and PEREIRA, 1965). Note the antenna-like fiber protruding from each vertex. This figure was kindly provided by Dr. Pereira
C. Electron micrograph of ad 5 virus contrasted with sodium silicotungstate showing the regular icosahedral symmetry of adenoviruses (HORNE et al., 1959)

have five neighbours and are called *pentons* (GINSBERG et al., 1966). Each penton unit consists of a penton base anchored in the capsid and a projection (VALENTINE and PEREIRA, 1965; NORRBY, 1966). The latter consists of a rod like portion with a knob attached at the distal end and is referred to as the *fiber*. Two types of hexons may be defined: (i) Those located in juxtaposition with the pentons, *peripentonal hexons*, which differ topologically from the remaining hexons since they have the penton base as one of their six neighbours; (ii) Those 180 which form the triangular facets and the edges of the icosahedron. The latter hexons may be released from the virions in aggregates of nine hexons (ninemers) which form a defined structure with 3-fold rotational symmetry as shown in Figure 1 A (SMITH et al., 1965; RUSSELL et al., 1967b; MAIZEL et al., 1968b; PRAGE et al., 1970; CROWTHER and FRANKLIN, 1972; ISHIBASHI and MAIZEL, 1974a). Inside the capsid there is a core, which contains the DNA and additional proteins as originally revealed by electron microscopy of thin sections of particles stained with uranyl acetate. This core has a diameter of 40—45 nm and its structure is destroyed by both DNase and trypsin (EPSTEIN 1959; EPSTEIN et al., 1960; BERNHARD et al., 1961). Isolated core structures have also been studied in the electron microscope by negative staining after disintegration of the virus with heat (RUSSELL et al., 1967b) or formamide (STASNY et al., 1968); a mesh-like unorganized morphology has usually been observed (RUSSELL et al., 1967b; STASNY et al., 1968; LAVER et al., 1968). The core contains all the DNA and about 20 per cent of the total protein of the virion. Acid extraction (PRAGE et al., 1968, 1970; RUSSELL et al., 1968), treatment with 6 M lithium iodide (NEURATH et al., 1970b) or SDS (MAIZEL et al., 1968a) releases all the core proteins from the DNA. ROBINSON et al. (1973) have reported that a circular protein-DNA complex is released after degradation of virus particles with guanidine-HCl. Subsequent phenol extraction linearized the DNA, which suggests that inside the virus particle the two termini of the adenovirus DNA are linked to each other with a protein.

III. The Composition of the Virion

The ad2 virion which has been studied in detail, has an estimated particle weight of 175×10^6 daltons (GREEN et al., 1967a). It contains 13 per cent DNA corresponding to 23×10^6 daltons of DNA (GREEN et al., 1967b; VAN DER EB et al., 1969) and the adenovirion is composed only of DNA and protein. Unlike most other viruses the adenoviruses have capsid units which are soluble in nondenaturing solvents. This has greatly facilitated the purification and characterization of the virus components and made it possible to develop methods for sequential disintegration of the virion (MAIZEL et al., 1968b; LAVER et al., 1969; PRAGE et al., 1970; EVERITT et al., 1973). The pentons alone or together with peripentonal hexons can be selectively removed by dialysis against distilled water (LAVER et al., 1969) or Tris-maleate buffer pH 6.0—6.5 (PRAGE et al., 1970). The release of pentons is accompanied by the release of additional antigens probably located in the peripentonal region. After treatment of the virus particle with SDS, urea or pyridine (SMITH et al., 1965; MAIZEL et al., 1968b; PRAGE et al., 1970) the capsid is disrupted and the hexons from the triangular facets are released as groups of nine hexons or ninemers (Fig. 1A).

The polypeptide composition of adenoviruses has been studied by SDS polyacrylamide electrophoresis. MAIZEL et al. (1968a, b) showed that the virion of adenovirus types 2, 7 and 12 contains a minimum of 9 polypeptides which range

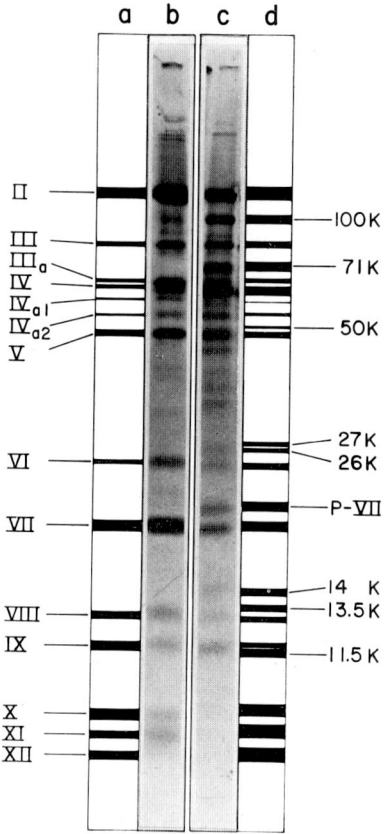

Fig. 2. Virion and virus-induced polypeptides in adenovirus infected cells. SDS-polyacrylamide gel autoradiograms of ^{35}S-methionine-labeled purified virus (b) and an extract of infected cells (c). The gel contained 15 per cent acrylamide and 0.08 per cent bisacrylamide. The extract was obtained from cells which were labeled for 1 hour with ^{35}S-methionine 18 hours after infection and then chased for 12 hours with 30 µg/ml of unlabeled methionine. Frames a and d are drawings which indicate all virion polypeptides (a) and all 22 virus-induced polypeptides (d). The nomenclature of EVERITT et al. (1973) and ANDERSON et al. (1973) is used to designate polypeptide bands. The non-structural polypeptides in the cell extract are designated by their molecular weight $\times 10^{-3}$ (K). P VII has been shown to be a precursor of virion polypeptide VII (ANDERSON et al., 1973). The virus induced polypeptide 27 K appears to be the precursor to virion polypeptide VI (ANDERSON et al., 1973; ÖBERG et al., 1975), and polypeptide 26 K is probably the precursor to virion polypeptide VIII (ÖBERG et al., 1975). Reprinted by permission from ANDERSON et al., (1973)

in size from 7,500 to 120,000 daltons. With a high resolving SDS polyacrylamide gel technique as many as 15 polypeptides have been resolved from ad2 virions (MAIZEL, 1971; EVERITT et al., 1973; ANDERSON et al., 1973) as shown in Figure 2. Some polypeptides may be generated by proteolytic degradation (PEREIRA and

SKEHEL, 1971), and some may be precursor polypeptides (ISHIBASHI and MAIZEL, 1974a) but 8 of the polypeptides have so far been shown to be antigenically distinct and reside in different structures after sequential degradation of the virion (EVERITT et al., 1973; EVERITT and PHILIPSON, 1974). Five of the polypeptides are integral parts of the capsomeres or the core. Thus, polypeptide II is the only

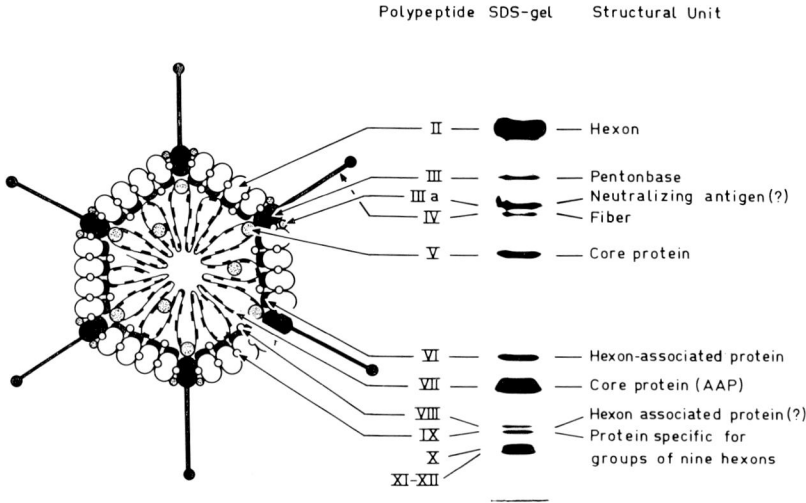

Fig. 3. A tentative model of the location of different proteins in the ad 2 virion. The core protein V may be located inside at the vertices since it is partially released with the peripentonal region (EVERITT et al., 1973). It has been estimated that protein VII may neutralize about 50 per cent of all phosphate residues in the DNA (PRAGE and PETTERSSON, 1971). The molar ratio between polypeptide VI and the hexon polypeptide is about 2 and the native protein VI exists as a dimer (EVERITT and PHILIPSON, 1974). Protein VI is not iodinated by lactoperoxidase (EVERITT, personal communication), which suggests that protein VI is located at the inner surface of the hexons. Peripentonal hexons also possess this protein. Protein IX appears to be the cementing substance between hexons from the facets, since it is associated with groups of nine hexons (EVERITT et al., 1973). Protein VIII is also associated with the hexons and may reside at the inner surface of the triangular facets since it is not enzymatically iodinated in intact virions (EVERITT, unpublished). Polypeptide IIIa has been proposed to be located at the peripentonal region. The localization of protein X is unknown. The polypeptide composition of the virion proteins as identified in a stained exponential (10—16 per cent) SDS-polyacrylamide gel are also shown. This figure was kindly provided by Dr. E. Everitt

polypeptide which is detected in purified hexons from infected cells. Polypeptide III resides in the penton base, IV in the fiber and polypeptides V and VII are associated with the core. The remaining polypeptides have been tentatively located in the virion as indicated in Figure 3. Polypeptide VI can be demonstrated in all fractions from disintegrated virus which contain hexons and sediments with hexons in sucrose gradients at low salt concentration irrespective of whether the hexons are obtained after disintegration of virions by freezing and thawing or pyridine treatment (EVERITT et al., 1973). Polypeptide VI is therefore probably associated with the hexons in the virion. Polypeptide VI cannot be iodinated by lactoperoxidase when intact virions are labeled (unpublished), which may suggest that it is located internal in the capsid as indicated in Figure 3. There appear to be about two molecules of polypeptide VI per molecule of hexon and this polypeptide seems to occur as a dimer in the native state (EVERITT and PHILIPSON, 1974). Polypeptide VIII is recovered in association with hexons after freezing and thawing but not after degradation with pyridine. Thus, it is possible that this polypeptide is associated with hexons in the capsid but that pyridine dissociates the complex (MAIZEL et al., 1968b; EVERITT et al., 1973). Polypeptide IX is present in ninemers of hexons (MAIZEL et al., 1968b; EVERITT et al., 1973) but is not recovered in association with hexons when the capsid is disintegrated into single capsomeres by freezing and thawing. Polypeptide IIIa is released from the virion together with the peripentonal hexons after dialysis against Tris-maleate buffer, pH 6.3 or after pyridine treatment, and may therefore be located in the peripentonal region (EVERITT et al., 1973). The origin of the smallest polypeptide X is not yet established and this band has been resolved into 3 components — X, XI and XII (Fig. 2) (ANDERSON et al., 1973). In addition to the polypeptides discussed above several minor components (polypeptides IVa_1, IVa_2, Va, Vb, VIa, VIb and VIIIa), which each constitute less than 0.1 per cent of the total mass of protein in the virion, can be observed in preparations of purified virions (EVERITT et al., 1973; ANDERSON et al., 1973; ISHIBASHI and MAIZEL, 1974a; see also Fig. 2). They are present in amounts which would correspond to only a few copies per virus particle and may not represent polypeptides in the mature virion (EVERITT et al., 1973). ISHIBASHI and MAIZEL (1974a) have proposed that purified preparations of adenovirus contain two populations of particles with the same buoyant density. One consists of mature virions and the other of so called young virions. "Young virions" contain five polypeptides (Va, Vb, VIa, VIb, VIIIa), which are absent in mature particles. During virus maturation these five polypeptides are presumed to be cleaved into stable polypeptides. It has been possible to demonstrate polypeptide cleavage *in vitro* after incubation of pulse-labeled virions with extracts from infected cells (ISHIBASHI and MAIZEL, 1974a). Finally, it should be pointed out that we so far lack positive evidence that all polypeptides of mature virions have unique primary structures.

Several reports describe virus-specific antigens of unknown origin from disrupted virions or in extracts from adenovirus infected cells. BERMAN and ROWE (1965) described a "D" antigen in cells infected with ad12. It is possible that this antigen is identical to the free penton bases which later have been shown to be present in cells infected with ad12 (NORRBY and ANKERST, 1969). PEDERSEN and GINSBERG (1967) identified by immunoelectrophoresis an antigen from disintegrated ad5 virions which was shown to be distinct from the capsid antigens. RUSSELL et al. (1967a) reported on the presence of another antigen, the P-antigen. This antigen was detected by antisera prepared against extracts of infected cells in which DNA synthesis was inhibited with cytosine arabinoside. It is produced both early and late after infection with ad5 and RUSSELL and KNIGHT (1967) showed that disrupted virions in contrast to intact virus particles reacted with antisera against the P-antigen. This suggests that one of the P-antigen components is an internal protein of the virion. Since the P-antigen is heat labile and accumulates in the nucleus without requirement for DNA synthesis, it has also been claimed to be related to the adenovirus T-antigen (see section V:G:1; RUSSELL and KNIGHT, 1967; HAYASHI and RUSSELL, 1968). Thus, it seems likely that P-antisera react with several antigens including T-antigen and an internal component of the virion.

IV. The Adenovirus Genome

A. Physical-Chemical Properties

The adenovirion contains one linear molecule of duplex DNA which has no single strand breaks. The molecular weight of DNA from different human serotypes has been determined to be 20 to 23×10^6 (GREEN et al., 1967b; VAN DER EB et al., 1969). The partial denaturation maps of DNA from ad2, ad5 and ad12 are unique (DOERFLER and KLEINSCHMIDT, 1970; DOERFLER et al., 1972; ELLENS et al., 1974) but the maps of the former two viruses are similar. In addition it has been shown that terminal fragments of adenovirus DNA are unique (MURRAY and GREEN, 1973) which together with the partial denaturation patterns show that the adenovirus DNA is not circularly permuted. Digestion of the adenovirus DNA with exonuclease III does not generate circular molecules (GREEN et al., 1967b) indicating that adenovirus DNA lacks terminal redundancies. Instead, the adenovirus DNA has a novel property in that denaturation and renaturation of intact DNA molecules at low concentrations generates single stranded circles (GARON et al., 1972; WOLFSON and DRESSLER, 1972). Since the majority of unit length single strands can be recovered as rings, both strands must be able to form circles. The circles are opened by digestion with exonuclease III which shows that they are maintained by hydrogen bonds. The most likely interpretation of these findings is that each strand of the adenovirus DNA contains inverted terminal repetitions, which form a circle which is held together by a panhandle-like structure. A terminal inverted repetition has so far only been observed in double stranded adenovirus DNA and in the single stranded DNA from adenovirus associated virus (AAV) — a member of the parvovirus group (KOCZOT et al., 1973). The function of this inverted terminal repetition is unclear. Since a protein seems to circularize double stranded adenovirus DNA (ROBINSON et al., 1973) it is conceivable that the identical sequences at the two ends of the viral DNA are recognition sites for this protein.

Usually single stranded DNA is not retained by hydroxylapatite in 0.12 M phosphate at 65° C. However, denatured adenovirus DNA is retained by hydroxylapatite under the same conditions (TIBBETTS et al., 1973). The retention is dependent on the extent of fragmentation of the adenovirus DNA and the fraction retained decreases slowly until the length is reduced to about 10 per cent of the intact DNA. With more extensive degradation the fraction retained decreases drastically. The formation of single strand circles may contribute to the retention of intact single stranded adenovirus DNA by hydroxylapatite, but the described behaviour of the fragmented DNA points at an additional feature of adenovirus DNA. Presumably several short complementary or partially complementary sequences are dispersed throughout the single strands of the adenovirus DNA. Figure 4 shows schematically these structures in adenovirus DNA.

The complementary strands of adenovirus DNA have been separated by density gradient centrifugation in the presence of ribocopolymers such as poly (I, C) and poly (U, G) (KUBINSKI and ROSE, 1967; LANDGRAF-LEURS and GREEN, 1971; PATCH et al., 1972), but the separation has proved difficult in practice. Not only do the conditions for the copolymer binding seem to be important but also the quality of the copolymer used. It is the experience of several investigators that many commercially available preparations of polymers are inefficient in

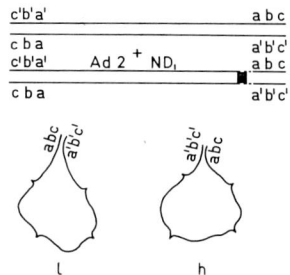

Fig. 4. Schematic drawing of some characteristic structures of ad 2 DNA. The adenovirus DNA contains an inverted terminal repetition (top figure, WOLFSON and DRESSLER, 1972; GARON et al., 1972). The genome of the non-defective adeno-SV 40 hybrid virus, ad 2^+ND_1, which has been used to map several restriction endonuclease fragments, is also shown. About 240,000 daltons of SV 40 DNA is inserted near the right hand end of the hybrid DNA where about 1.3×10^6 daltons of ad 2 DNA has been deleted (KELLY and LEWIS, 1973, see Table 6 and Fig. 17). The circles of the single stranded adenovirus DNA which are formed because of the inverted terminal repetition are illustrated for both the l- and h-strand at the bottom of the figure. The single strands appear to contain regions with intramolecular complementarity and there appear to be 4—8 such regions per genome (TIBBETTS et al., 1973)

separating the adenovirus strands, but that occasional batches can give reproducible and near complete separation (TIBBETTS et al., 1974). The reason for this variability is not known. Separated adenovirus strands have been used for the characterization of RNA synthesis and processing in productive infection as will be discussed in a separate section V: D.

After controlled shearing of ad 2 DNA, double stranded half molecules can be separated by mercury-CsCl gradient centrifugation (KIMES and GREEN, 1970; DOERFLER and KLEINSCHMIDT, 1970). Endonuclease EcoRI (a restriction enzyme from E. coli carrying a drug resistant transfer factor RTF-I), cleaves ad 2 DNA into six unique double stranded fragments which can be separated by gel electrophoresis (PETTERSSON et al., 1973). They have been designated A-F and their order has been found to be A B F D E C. The ordering of the fragments was achieved by electron microscopic analyses of partially denatured fragments (DELIUS, personal communication) and of heteroduplex molecules formed between EcoRI fragment strands and strands from other adenoviruses (SHARP et al., 1974b). The A fragment which accounts for 59 per cent of the genome

can be further cleaved into five fragments by one restriction endonuclease from
H. parainfluenzae (HpaI) (SHARP et al., 1973; GALLIMORE et al.; 1974). The EcoRI
fragments have already proved useful for the mapping of early and late genes

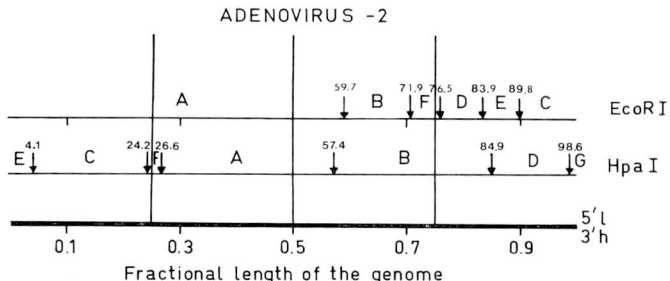

Fig. 5. Cleavage sites on ad 2 DNA of two different bacterial restriction endonucleases.
The endo R. EcoRI enzyme is obtained from a strain of E. coli (RY 13) carrying the
drug resistance transfer factor RTF-1. The size of the fragments was determined by
polyacrylamide gel electrophoresis and electron microscopy (PETTERSSON, et al. 1973,
and Table 1) and the order of the fragments was determined by electron microscopy
(MULDER et al., 1974 b). The endo R. Hpa I is one of two enzymes isolated from
Hemophilus parainfluenzae (SHARP et al., 1973). The size and the order of the fragments
was determined by MULDER et al., 1974 b

Table 1. *Molecular Weights (Megadaltons) of endo R · Eco RI Fragments of DNA of Various Adenoviruses*[a]

	Intact	A	B	C	D	E	F
Ad 2	22.9	13.6	2.8	2.3	1.7	1.4	1.1
Ad 2+ND$_1$	22.2	13.5	2.8	2.4	2.5 (C′)[b]	—	1.1
Ad 5	22.9	17.6	3.7	1.7			
Ad 3	22.5	19.2	2.9	0.4			
Ad 7 (E 46-LL)	22.6	19.3	2.9	0.3			
Ad 7 (cl 19)	22.3	19.4	3.0	—			
Ad 12	21.5	7.7	5.7	3.5	2.6	1.6	0.4

[a] The values in this table were estimated by electron microscopy and gel electrophoresis. Data from MULDER et al., 1974 a.
[b] This fragment, referred to as C′, contains the portion of SV 40 DNA which is included in the hybrid DNA.

on the viral chromosome (TIBBETTS and PETTERSSON, 1974; PETTERSSON and
PHILIPSON, 1974; PETTERSSON et al., 1975; SHARP et al., 1974 a). Ad 5, ad 3 and
ad 7 (clone E 46-LL) are cleaved in 3 and ad 12 in 6 fragments by the EcoRI
endonuclease (Table 1, MULDER et al., 1974 a). Figure 5 shows schematically the
structure of ad 2 DNA and the cleavage sites for different restriction endonucleases.

B. Homology between DNA from Different Serotypes

Based on oncogenicity the human adenoviruses have been divided into four
subgroups A, B, C and D (HUEBNER, 1967; MCALLISTER et al., 1969 b). For the
human adenoviruses a correlation has been established between oncogenicity and

content of guanosine and cytosine in the DNA (Piña and Green, 1965). The highly oncogenic adenoviruses (subgroup A) have a GC content of 48—49 per cent, the weakly oncogenic adenoviruses (subgroup B) have a GC content of 49—52 per cent, and the so called non-oncogenic viruses (subgroups C and D) have a GC content of 57—59 per cent (Piña and Green, 1965; Green, 1970). Simian adenoviruses do not follow this rule, since for instance, the highly oncogenic SA7 has a GC content of around 60 per cent (Piña and Green, 1968; Goodheart, 1971). It has been shown by reciprocal DNA-DNA and DNA-RNA hybridizations that the members of one subgroup are more closely related to each other than to

Table 2. *Homology between Different Adenovirus DNAs as Estimated by Filter Hybridization and Electron Microscopical Heteroduplex Mapping*

Subgroups represented in DNA pairs	Homology by			
	Filter hybridization[a]		Heteroduplex mapping[b]	
	Per cent	Serotypes	Per cent	Serotypes
A × A	80	(ad 12, ad 18, ad 31)	15	(ad 12, ad 31)
B × B	85	(ad 3, ad 7, ad 11, ad 14, ad 16)	88	(ad 3, ad 7)
C × C	90	(ad 1, ad 2, ad 5, ad 6)	85	(ad 1, ad 2, ad 5)
A × B	10—25	(all crosses)	—	
A × C	10—25	(ad 12 × ad 1, ad 2, ad 5, ad 6)	—	
B × C	25	(ad 7 × ad 1, ad 2, ad 5, ad 6)	10	(ad 7 × ad 2)
Simian × A	10	(SA 7 × Ad 12)	—	
Simian × B	20	(SA 7 × Ad 7)	—	
Simian × C	23	(SA 7 × Ad 2)	—	

[a] Modified from Green, 1970. Data from Fujinaga et al., 1969; Lacy and Green, 1964, 1965, 1967; Piña and Green, 1968.
[b] Data from Garon et al., 1973. Heteroduplex molecules were examined by electron microscopy after spreading DNA at 85 per cent formamide. At 50 per cent formamide the cross within subgroup A showed 85 per cent homology.

members of the other groups (Table 2 and for a review see Green, 1970). A recent report (Garon et al., 1973) describes electron microscopic analysis of heteroduplexes between DNAs from different adenoviruses. This analysis has confirmed the existence of extensive homologies between DNAs from viruses within the same subgroup. Heteroduplexes formed between DNAs from serotypes which both belong to subgroups B or C contained around 85 per cent homologous sequences when the DNA was analyzed in the presence of 85 per cent formamide. Figure 6 shows the heteroduplexes formed between ad 1 and ad 5 at different concentrations of formamide. Analysis at lower formamide concentrations shows that the unmatched regions consist of both heterologous sequences and sequences that are partially homologous. Heteroduplexes formed with DNA from members of the highly oncogenic subgroup A (ad 12, ad 18 and ad 31) contain more pronounced heterologies and at 85 per cent formamide only 15 per cent of the heteroduplex of ad 12 and ad 31 DNA is homologous. The heteroduplexes formed between DNAs from subgroup B viruses (ad 3, ad 7 and ad 21) and subgroup C viruses (ad 1, ad 2 and ad 5) all exhibit relatively simple patterns with two short regions, which are not base-paired. These regions

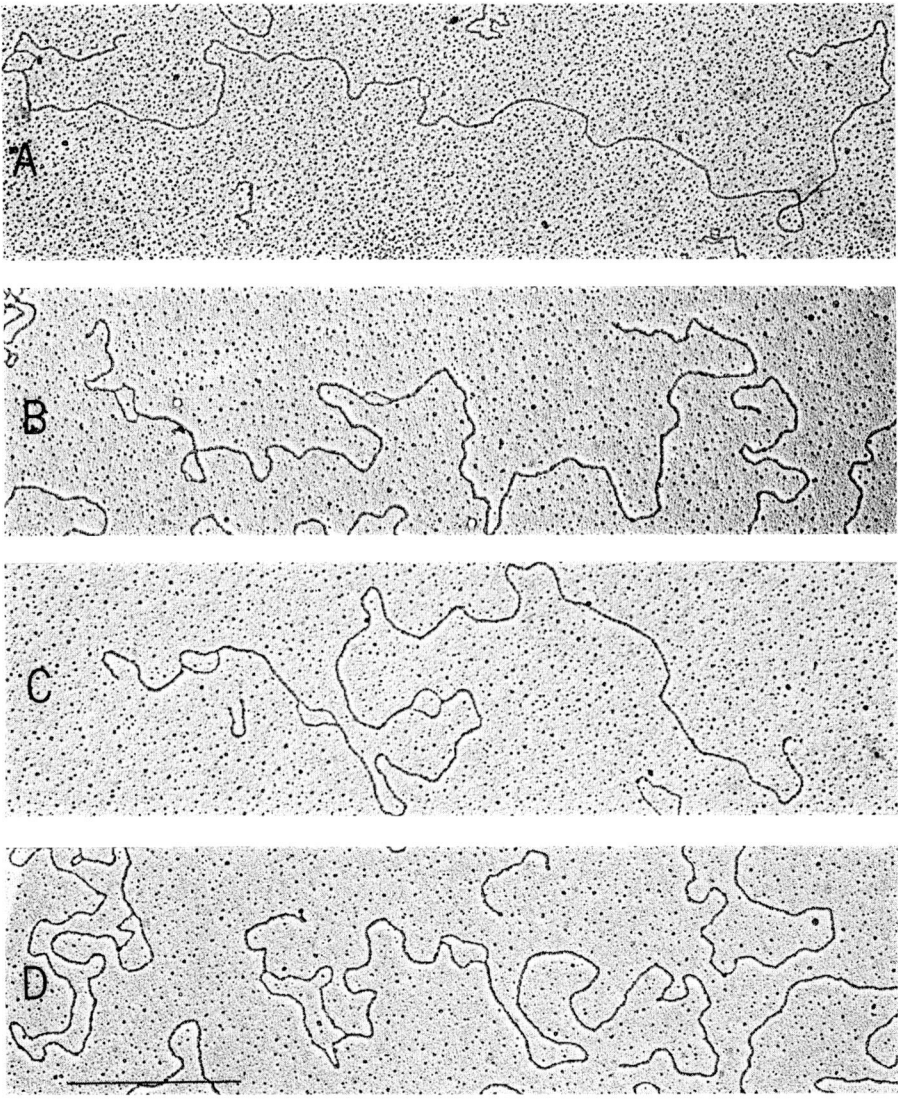

Fig. 6. Electron micrographs of heteroduplex molecules between DNA from ad 1 and ad 5.
The DNA was mounted by the formamide isodenaturing technique. Panels show heteroduplex molecules spread at formamide concentrations of (A) 50 per cent, (B) 60 per cent, (C) 70 per cent, and (D) 85 per cent. A similar heteroduplex between ad 1 and ad 2 was published by GARON et al. (1973) Bar represents 1 μm. This figure was kindly provided by Dr. C. F. GARON

appear limited to two areas of the molecules which with minor variations are located at 0.08—0.22 and 0.35—0.50 fractional lengths from one end of the molecule. The inserted piece of SV 40 DNA in the ad 7-SV 40 hybrid DNA (E46$^+$) (see section VII) is located at one of these regions (GARON et al., 1973). Quite complex patterns of unmatched regions are observed for heteroduplexes formed between different group A viruses, but at low formamide concentration the most pronounced heterologies always appear at the two positions indicated above. As pointed out by GARON et al. (1973), this would imply that there are two specific regions in adenovirus DNA, which are especially prone to genetic "drift", and it is possible that the sequences in these two regions correspond to the same genes in all serotypes. Heteroduplex molecules between DNA from different subgroups finally appear to contain only about 10 per cent homologous sequences.

C. Infectivity of Adenovirus DNA

Several years ago, DNA from simian adenovirus 7 (SA7) was reported to be infectious and tumorigenic (BURNETT and HARRINGTON, 1968a and b). Using the DEAE dextran technique NICOLSON and McALLISTER (1972) were able to show that human ad1 DNA also is infectious but the infectivity was extremely low. Recently GRAHAM and VAN DER EB (1973a and b) introduced a new technique to assay infectivity and transforming activity of adenovirus DNAs. The method utilizes the adsorbtion of the DNA to calcium phosphate precipitates, which facilitates the uptake of infectious DNA by the cell. The technique is efficient and reproducible and it was demonstrated that 1—10 plaque forming units and about one transformed focus can be observed per µg of ad5 DNA.

V. The Productive Infection

The lytic cycle of adenoviruses has primarily been studied in suspension cultures of KB and HeLa cells. The time course of the infection with ad2 is shown in Figure 7. When the cells are infected at high multiplicities (50—100 PFU/cell) a synchronous response is observed, where two functionally different phases can be distinguished: an early phase, which precedes viral DNA replication and a late phase beginning with the onset of viral DNA synthesis around 6 hours after infection. During the early phase about 40 per cent of the viral genome is expressed (TIBBETTS et al., 1974) and the so called T- and P-antigens are synthesized (GILEAD and GINSBERG, 1968a and b; RUSSELL and KNIGHT, 1967). Between 8 and 12 hours after infection, there is a dramatic change in the production of viral mRNA. About 90 per cent of the genome is now expressed (FUJINAGA and GREEN, 1970; TIBBETTS et al., 1974) and the amount of viral RNA made increases about 10-fold. The viral mRNA accumulates in the cytoplasm and at 14 hours post infection the synthesis of host cell proteins is to a large extent replaced by synthesis of viral products — mostly viral structural proteins (BELLO and GINSBERG, 1967; WHITE et al., 1969; RUSSELL and SKEHEL, 1972). New virus particles begin to appear at 15 hours and the infectious cycle for ad2 and ad5, both members of subgroup C, is completed within 20—25 hours. Similar time courses have been obtained with other serotypes (GREEN et al., 1970; SUNDQUIST et al., 1973b). However, the growth cycle of the highly oncogenic group A viruses is somewhat longer than for ad2 (MAK and GREEN, 1968) and when primary cells are used the growth cycle is prolonged at least 12—15 hours also for group C viruses (LEDINKO, 1970).

A. Early Interaction between Virus and Cell

The adenovirus eclipse has been studied primarily with ad 2 and ad 5. Two different techniques have been employed: (i) electron microscopy of thin sections of infected cells (DALES, 1962; MORGAN et al., 1969; CHARDONNET and DALES, 1970a and b; CHARDONNET and DALES, 1972) and (ii) analysis of the fate of radio-

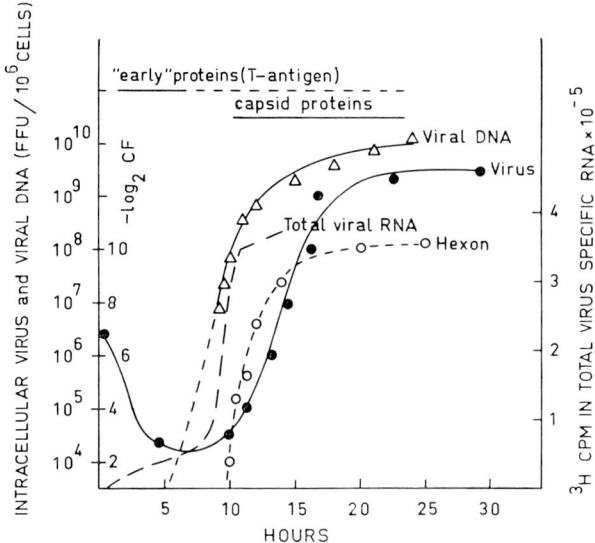

Fig. 7. Growth curve of ad 2 in suspension culture of KB cells. (●——●) intracellular virus measured as fluorescent focus forming units per 10^6 cells; (○——○) Hexon antigen monitored by complement fixation; (△——△) Synthesis of viral DNA from GREEN et al. (1970); (--------) Virus specific RNA during the infectious cycle measured as radioactivity exhaustively hybridizing to ad 2 DNA on filters. Reprinted from PHILIPSON and LINDBERG (1974)

actively labeled virus (LAWRENCE and GINSBERG, 1967; PHILIPSON, 1967; SUSSENBACH, 1967; PHILIPSON et al., 1968; LONBERG-HOLM and PHILIPSON, 1969). The ratio of particles to infectious units for serotypes ad 2 and ad 5 is in the order of 10—30, but for other serotypes the ratio can be as high as 2000 (GREEN et al., 1967a; LONBERG-HOLM and PHILIPSON, 1969). It should be pointed out therefore that the methods used for studying the early interaction only reveal the fate of the majority of the particles and that this might not reflect the true infectious pathway. In any case it appears that the virions are adsorbed to specific receptors of the cell surface. There are about 10^4 such receptors per KB cell (PHILIPSON et al., 1968). Uncoating of the virus occurs rapidly after infection and seems to involve three steps. The *first* step gives rise to particles, whose DNA is partially accessible to DNase digestion and biochemical evidence suggests that these particles are devoid of penton capsomeres (SUSSENBACH, 1967). This event occurs within a few minutes after attachment of the virus to the cell surface (SUSSENBACH, 1967; LONBERG-HOLM and PHILIPSON, 1969; MORGAN et al., 1969; BROWN and BURLINGHAM, 1973). Whether the uptake of the virus is brought about mainly by pinocytosis (DALES, 1962; CHARDONNET and DALES, 1970a and b) or by direct penetration of the plasma membrane (LONBERG-HOLM and PHILIPSON, 1969; MORGAN et al., 1969; BROWN and BURLINGHAM, 1973) is unresolved, and exactly where this first step of uncoating takes place is not known. It has been suggested that microtubules are engaged in the transfer of virions from the plasma membrane to the nuclear pores (CHARDONNET and DALES, 1972). The

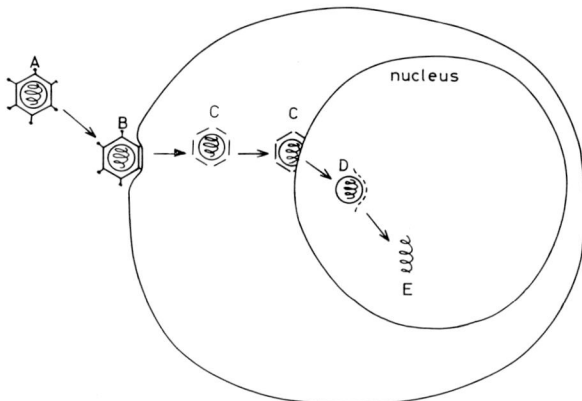

Fig. 8. Diagram of events during the first two hours of infection with adenovirus type 2. Intact virions (A) and particles attached to the plasma membrane (B) are shown. Particles which lack the peripentonal region (C) enter the cytoplasm and the cores are expelled in the nuclei (D) and finally converted in free DNA (E). Only DNA containing components are shown. Modified from LONBERG-HOLM and PHILIPSON (1969)

second step results in removal of the hexon capsomeres from the DNA-protein core. As expected, this intermediate is even more sensitive to DNase. Electron microscopy has shown that cores enter into the nucleus apparently leaving the capsid structure behind (MORGAN et al., 1969; CHARDONNET and DALES, 1972). It has been suggested that the transfer of the cores into the nucleus is an ATP-dependent process (DALES and CHARDONNET, 1973). The *third* and final step in the uncoating of the adenovirus DNA occurs in the nucleus of the cell during the second hour after infection and the final product of uncoating appears to be DNA, still intact, but free of virion proteins (LONBERG-HOLM and PHILIPSON 1969). Figure 8 shows a schematic pathway for the uncoating events of ad 2. The time course of the uncoating process is dependent on the multiplicity of infection, but at high multiplicities the overall time required is 1—2 hours (LONBERG-HOLM and PHILIPSON, 1969). The mechanism and cell structures involved in the different steps are unknown, but it should be pointed out that the three intermediates seen during the *in vivo* uncoating resemble structures which may be obtained by sequential disintegration of the virus (PRAGE et al., 1970). The uncoating proceeds normally in the presence of inhibitors of protein and nucleic acid synthesis, so it has to be brought about by preexisting enzymes either present in the cell or in the infecting virus (LAWRENCE and GINSBERG, 1967; PHILIPSON, 1967).

B. Early RNA Synthesis

Production of viral RNA starts immediately after the viral genome has reached the nucleus and at 5 hours after infection around 10—15 per cent of the newly synthesized mRNA associated with polyribosomes appears to be virus-coded (LINDBERG et al., 1972). The pattern of early viral mRNA is unchanged until the onset of viral DNA synthesis (PARSONS and GREEN, 1971; LUCAS and GINSBERG, 1971; WALL et al., 1972; LINDBERG et al., 1972).

The viral RNA molecules in the nucleus are as heterogenous as nuclear RNA (HnRNA) in uninfected cells and the largest viral RNA molecules appear to be between 5 and 10×10^6 in molecular weight (PARSONS and GREEN, 1971; WALL et al., 1972). However, viral RNA associated with the polyribosomes is smaller, and distinct mRNA peaks can be resolved by gel electrophoresis (PARSONS and GREEN, 1971; LINDBERG et al., 1972). These observations have led to the idea that adenovirus mRNA may be generated from high molecular weight precursors in the nucleus.

The viral RNA appears to be polyadenylated in the nucleus and since, poly (A) does not hybridize significantly to adenovirus DNA, it is concluded that polyadenylation is a posttranscriptional event (PHILIPSON et al., 1971). Poly (A) segments recovered from RNA, which has been selected by hybridization to virus DNA, has the same size as the poly (A) in mRNA from uninfected cells, i.e. 180—200 nucleotides (PHILIPSON et al., 1971). Because of its poly (A) content, polyribosomal RNA from adenovirus-infected cells has been fractionated by affinity chromatography on poly (U)-Sepharose or oligo (dT) cellulose (LINDBERG et al., 1972). The results of such experiments indicate that most if not all of early viral mRNA species contain poly (A). As in the uninfected cell, the adenosine analogue cordycepin (3′-deoxyadenosine) inhibits polyadenylation and in the presence of this drug no viral mRNA is transported to the cytoplasm (PHILIPSON et al., 1971). Thus, it has been concluded that polyadenylation is necessary for proper processing and transport of adenovirus mRNA to the cytoplasm (DARNELL et al., 1971 b).

Three major size classes of early viral mRNA have been found (PARSONS and GREEN, 1971; LINDBERG et al., 1972): one class sediments at around 15S and has an approximate molecular weight of $0.3—0.4 \times 10^6$, a second class sediments at around 20S with an approximate molecular weight of 0.8×10^6, and a third heterogenous population ranges from 20—40S (molecular weight $1.0—3.5 \times 10^6$) (Fig. 9). Each of these three classes may contain several species of mRNA and it has recently been established that the 20S class contains at least two different species of mRNA derived from the h- and l-strands of the viral DNA, respectively (CRAIG et al., 1974; PHILIPSON et al., 1974). The fraction of the viral genome which is expressed early and late after infection will be discussed in a separate section (section V: D).

The transcription of the adenovirus genome early after infection is not sensitive to inhibitors of protein synthesis (PARSONS and GREEN, 1971; WALL et al., 1972). Since no RNA polymerase activity has been detected in adenovirions, this suggests that early viral RNA is synthesized by a cellular enzyme. In accordance with this, several investigators (LEDINKO, 1971; PRICE and PENMAN, 1972a; WALLACE and KATES, 1972; CHARDONNET et al., 1971) have found that viral

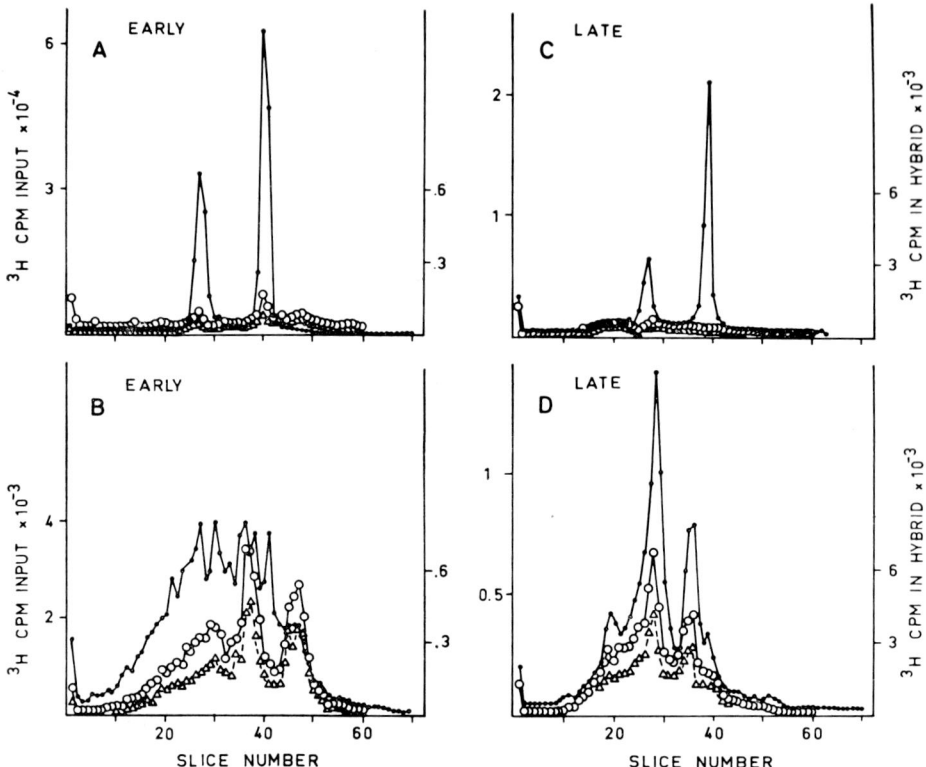

Fig. 9. Size distribution of early and late adenovirus mRNA.
Early and late polyribosomal RNA was fractionated by poly(U) Sepharose-chromatography. RNA lacking poly (A) (PUS I) and the fraction retained by the column (PUS II) was analyzed by gel electrophoresis (LINDBERG et al., 1972).
Panel A and C: early and late PUS I RNA, respectively.
Panel B and D: early and late PUS II RNA, respectively.
Total ^{3}H-cpm. ●────●, ^{3}H-cpm in viral mRNA before o────o, and after △─────△ RNase digestion of hybrids. Reprinted from LINDBERG et al. (1972)

RNA is transcribed by an enzyme which resembles one of the RNA polymerases of the host cell [polymerase II according to ROEDER and RUTTER (1970) which corresponds to polymerase B according to the nomenclature of CHAMBON et al. (1970)]. It is inhibited by α-amanitine and stimulated by increasing ionic strength. The same enzyme seems to be responsible for the major part of the adenovirus transcription both early and late after the infection. In the uninfected cell this enzyme appears to be responsible for the synthesis of heterogeneous nuclear RNA (HnRNA) (ZYLBER and PENMAN, 1971).

C. Late RNA Synthesis

The switch from early to late gene expression of the virus encompasses both a quantitative and a qualitative change in the synthesis of viral mRNA. Large amounts of viral mRNA accumulate in the cytoplasm: about 10 times more than in the early phase (GREEN et al., 1970; PHILIPSON et al., 1973). At 14 hours after infection newly synthesized mRNA exported to the cytoplasm is almost exclusively virus-specific and new classes of viral mRNA can be detected by gel electrophoresis or sucrose gradient centrifugation. Apparently a large part of the host mRNA on the polysomes is replaced by viral mRNA since the major part

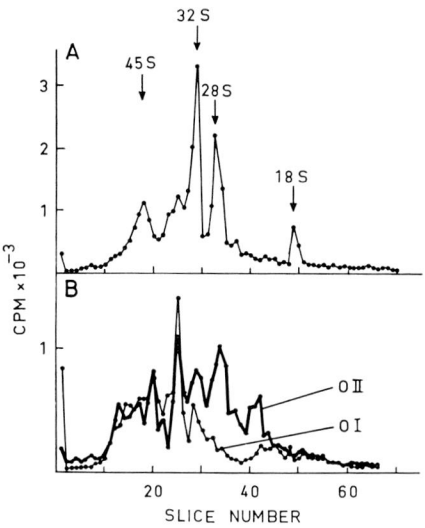

Fig. 10. Polyacrylamide gel electrophoresis of nuclear RNA late in adenovirus infection. KB cells infected with ad 2 were labeled with ^{3}H-uridine between 14—16 hours after infection. The nuclear RNA was extracted by phenol according to HOLMES and BONNER (1973) and fractionated on oligo (dT) cellulose according to AVIV and LEDER (1972). Panel A shows the total radioactivity in the fraction not retained by oligo (dT) cellulose and Panel B shows the virus specific RNA both in fractions with (O II) and without (O I) poly (A)

of protein synthesis now gives rise to what appears to be virus-coded proteins (BELLO and GINSBERG, 1967; WHITE et al., 1969; RUSSELL and SKEHEL, 1972). The transition to the late phase is prevented both by inhibitors of protein (GREEN et al., 1970) and DNA synthesis (FLANAGAN and GINSBERG, 1962).

Also late after adenovirus infection, a fraction of viral RNA in the nucleus is recovered as large RNA molecules with molecular weights between $5-10 \times 10^6$ (GREEN et al., 1970; McGUIRE et al., 1972; WALL et al., 1972). Like in the early phase, much larger viral RNA molecules are found in the nucleus than are recovered from the cytoplasm, implying that also late viral mRNA is derived by cleavage from larger precursors in the nucleus (Fig. 10). Evidence in favor of this interpretation was obtained by the use of the drug toyocamycin (an adenosine analogue) (McGUIRE et al., 1972). When this compound is applied to adenovirus-infected cells at moderate concentrations (1.75 µg/ml), giant viral RNA accumulates in the nucleus whereas no messenger RNA can be detected in the cytoplasm. The drug has an analogous effect on the synthesis of ribosomal RNA (TAVITIAN et al., 1968). Here the 45S precursor to ribosomal RNA is made, but fails to undergo cleavage to 28 and 18S ribosomal RNA.

Analysis of nuclear RNA purified by hybridization to viral DNA has demonstrated that also in the late phase viral RNA molecules goes through a step of posttranscriptional polyadenylation (PHILIPSON et al., 1971). Analysis of the late polyribosomal RNA after fractionation on poly (U) Sepharose has demonstrated that most if not all the viral mRNA contains poly (A) (LINDBERG et al., 1972) and that polysomal mRNA separates into two major viral mRNA classes; one class sediments at 22S and the other at 26S with approximate molecular weights of 0.9×10^6 and 1.5×10^6, respectively. In addition, several minor mRNA peaks of higher and lower molecular weight (LINDBERG et al., 1972; BHADURI et al., 1972) can be detected (Fig. 9).

In addition to viral mRNA a small molecular weight virus-specific RNA is synthesized in large amounts late after lytic infection. This RNA which sediments at 5.5S RNA is known as the "virus-associated RNA" (VA RNA). It contains 156 nucleotides and its sequence has been determined (OHE and WEISSMAN, 1971; OHE et al., 1969). It is synthesized in larger amounts than any viral mRNA and apparently by an enzyme which is different from that which is involved in mRNA synthesis (PRICE and PENMAN, 1972b). VA RNA is transcribed from the adenovirus DNA and there seems to be only one gene for VA RNA in the adenovirus chromosome (OHE, 1972). No functional role has as yet been ascribed to this RNA, although kinetic experiments suggest that VA RNA is intermittently associated with polyribosomes shortly after transport to the cytoplasm (BAUM and FOX, 1974; PHILIPSON unpublished).

D. Mapping of Early and Late Viral RNA on Adenovirus DNA

The complementary strands of adenovirus DNA can be separated with the ribocopolymer binding technique of SZYBALSKI et al. (1971). This was first shown by KUBINSKI and ROSE (1967) and LANDGRAF-LEURS and GREEN (1971) who separated the complementary strands of several adenovirus serotypes by CsCl gradient centrifugation in the presence of poly (I, C) or poly (U, G). LANDGRAF-LEURS and GREEN (1973) and PATCH et al. (1972) used this method to prepare separated strands which were immobilized on filters for hybridization studies. The results show that most of the RNA synthesized late in infection hybridizes to the l-strand (defined as the strand which has a lower buoyant density in CsCl when complexed to poly [U, G]) and that early RNA hybridizes to both the complementary strands (LANDGRAF-LEURS and GREEN, 1973; PATCH et al., 1972). Recently, TIBBETTS et al. (1974) used separated strands of ad2 DNA in liquid phase hybridization experiments to determine the fraction of each strand which hybridizes to messenger RNA early and late after infection. The results showed that early messenger RNA hybridizes to 25—30 per cent of the l-strand and 10—15 per cent of the h-strand. Late messenger RNA was found to be derived from 65—70 per cent of the l-strand and 25 per cent of the h-strand, and all sequences present in early messenger RNA could also be detected late after infection (TIBBETTS et al., 1974). This does not necessarily mean that all early sequences are synthesized late, and LUCAS and GINSBERG (1971) have indeed furnished evidence by competitive filter hybridization that some early mRNA is not synthesized late.

In order to localize the regions on the adenovirus genome which specifically hybridize to early and late RNA, hybridization experiments have been performed with the six endo R. Eco RI fragments of ad 2 DNA. SAMBROOK et al., (1974) have performed experiments designed to indicate whether individual cleavage fragments of ad 2 DNA contained regions of one or both complements represented in late mRNA. The results showed that late mRNA is derived from both strands of fragments A, B and C but only from one strand of fragments D, E and F. TIBBETTS and PETTERSSON (1974) and SHARP et al. (1974a) separated the complementary strands of all six endo R. Eco RI fragments of ad 2 DNA and hybrid-

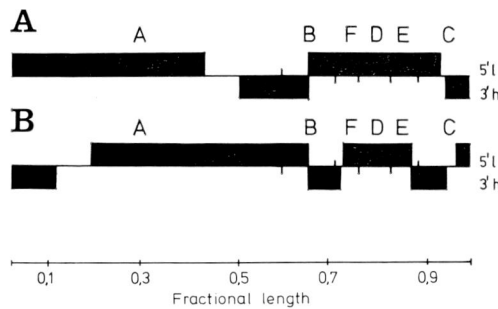

Fig. 11. Distribution of late viral mRNA sequences on ad 2 DNA.
Late cytoplasmic RNA was hybridized to separated strands of the six Eco RI fragments of ad 2 DNA (Table 3). The data were used to construct maps for the distribution of late mRNA sequences with reference to the complementary strands at the cleavage sites of endonuclease Eco RI. The data do not distinguish between the two possible alternatives A and B. The assignment of the 5′ and 3′ ends of each complement results from knowledge of the orientation of the integrated portion of SV 40 in ad 2+ND₁ DNA (MORROW and BERG, 1972) and the polarity of transcription of SV 40 with respect to endo R. Hin and endo R. Hpa cleavage maps (KHOURY et al., 1973; SAMBROOK et al., 1973) as well as a direct approach to determine the polarity (SHARP et al., 1974 a). Reprinted from TIBBETTS and PETTERSSON (1974)

ization experiments were performed with early and late messenger RNA in RNA excess. The results showed, like those of SAMBROOK et al. (1974), that late messenger RNA was derived from both strands of fragments A, B and C and predominantly from the l-strand of fragments D, E and F (Table 3). Knowing the order of the endo R. Eco RI fragments (SHARP et al., 1974b) it is possible to construct maps which show the regions on the two strands which are complementary to late messenger RNA. The least complicated map, which could be constructed with available data, indicates that late messenger RNA is derived from at least two separate regions on each of the complementary strands (Fig. 11). Comparisons

Table 3. *Hybridization of Early and Late Messenger RNA and Late Nuclear RNA with ^{32}P-Labeled Complement Specific DNA from the Six Endo R. Eco RI Cleavage Fragments of ad 2 DNA*

Fragment strand	Per cent DNA probe in hybrid with		
	Early messenger[a] RNA	Late messenger[a] RNA	Late nuclear[b] RNA
RI-Ah	9	15	13
RI-Al	21	71	84
RI-Bh	42	40	15
RI-Bl	7	52	88
RI-Ch	60	50	19
RI-Cl	9	31	87
RI-Dh	3	5	3
RI-Dl	72	82	91
RI-Eh	4	12	3
RI-El	40	82	88
RI-Fh	13	11	7
RI-Fl	23	84	87

[a] Data from PETTERSSON et al., 1975.
[b] Data from PETTERSSON and PHILIPSON, 1974.

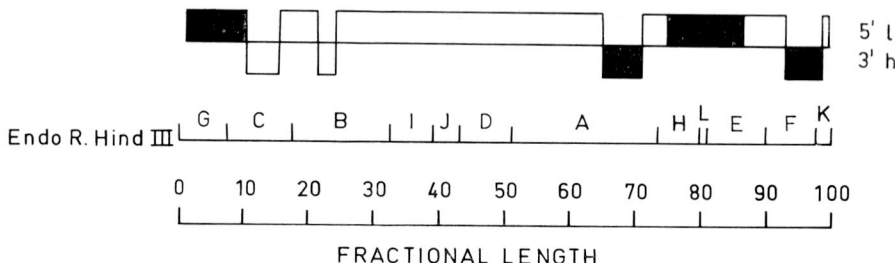

Fig. 12. A map of the early and late regions on the ad 2 chromosome. Cytoplasmic poly (A) containing RNA was hybridized to separated strands of the six endo R. *Eco* RI fragments and the results are summarized in Table 3. Knowing the order of the fragments (SHARP et al., 1974b), the least complicated maps for early and late mRNA were constructed as shown in Fig. 11. The maps do not take into account the small fractions of fragments *Eco* RI-Bl and *Eco* RI-Cl, which hybridize to early cytoplasmic RNA (Table 3).
In order to further resolve the left and right hand part of the map, hybridization experiments were performed with late cytoplasmic RNA and separated strands of endo R. *Hpa* I fragments and endo R *Hind* III fragments. The late mRNA map is represented by white and black bars and the early regions are represented by the black bars. There may still be some uncertainty concerning the RNA sequences in the Hind III B and C fragments. The map of the endo R Hind III fragments was kindly provided by Dr. R. J. Roberts of the Cold Spring Harbor Laboratory, New York.
Reprinted from PETTERSSON et al., 1975

of hybridization experiments with separated strands of the endo R. *Eco* RI fragments and early and late messenger RNA (SHARP et al., 1974 a; PHILIPSON et al., 1974; PETTERSSON et al., 1975) have shown that messenger RNA sequences which are present exclusively late are predominantly derived from the l-strand of fragments A, C, B and F and the h-strand of fragment A (Table 3). Additional fragments were obtained by a restriction endonuclease from *Hemophilus parainfluenzae* (HpaI) which generates 5 fragments from the large A fragment from endo R. *Eco* RI cleavage (Fig. 5). When these fragments were used to map early and late mRNA, a more detailed map of early and late mRNA sequences was established (SHARP et al., 1974 a; PHILIPSON et al., 1974; PETTERSSON et al., 1975) (Fig. 12). As indicated in Figure 12, early and late genes seem to be scattered in several locations on both complementary strands.

Several investigators have shown by hybridization competition experiments that some RNA sequences which are present in the nucleus never enter the cytoplasm (WALL et al., 1972; McGUIRE et al., 1972; LUCAS and GINSBERG, 1972a). Hybridization studies with separated strands indicate that late nuclear RNA hybridizes to about 85 per cent of the l-strand, whereas late cytoplasmic RNA is derived from only 65—70 per cent of the l-strand (PETTERSSON and PHILIPSON, 1974). Further hybridization studies with separated strands of the six endo R. *Eco* RI fragments indicate that late nuclear RNA hybridizes to 85—90 per cent of the l-strand of all these fragments. Since late messenger RNA is complementary to 40—50 per cent of the h-strand of both fragments B and C, this finding indicates that certain regions of the adenovirus chromosome are transcribed into complementary RNA sequences. Recently it has been established that isolated double stranded RNA late in adenovirus infection contains sequences complementary to at least 60 per cent of the h- and l-strand of ad2 DNA (PETTERSSON and PHILIPSON, 1974).

Studies on RNA synthesis in adenovirus-infected cells leave many problems unsolved. Nothing is yet known about initiation and termination of RNA synthesis *in vivo*. Experiments *in vitro* with *E. coli* RNA polymerase and ad2 DNA have shown that this enzyme initiates RNA synthesis at 5—10 preferred sites (PETTERSSON et al., 1974). On the other hand, nuclear RNA from adenovirus infected cells seems to contain RNA species which correspond in length to 50 per cent or more of one adenovirus DNA strand (PHILIPSON et al., 1974). This would indicate that the adenovirus DNA is transcribed into giant RNA molecules which are processed to messenger RNA. At least parts of the genome is symmetrically transcribed since complementary RNA sequences can be detected during adenovirus infection. Also, there is no clue to how the switch from early to late transcription is accomplished. Several alternatives seem conceivable: 1) The histone-like core proteins may impose restrictions on transcription early after infection. 2) The RNA polymerase may change its specificity during infection because of virus-induced control elements. 3) The adenovirus DNA may become integrated into the host cell genome even during productive infection and thereby gain access to additional promotor- and termination sites. 4) The switch to late transcription could be caused by changes in enzymes which process RNA.

Available information does not permit a choice between these alternatives but the findings that nuclear RNA both early and late after infection contains large RNA molecules, the size of which correspond to at least half of one strand (McGUIRE et al., 1972; WALL et al., 1972) favour the last alternative.

E. A Comparison between Cellular and Viral mRNA Production

In comparing recent results concerning the manufacturing of mRNA in mammalian cells in general with the corresponding findings in the adenovirus system, clear similarities as well as differences stand out. There is evidence suggesting that mRNA in animal cells is generated by cleavage of high molecular weight precursors in the nucleus and that the precursors are part of the metabolically active pool of RNA molecules called heterogeneous nuclear RNA, HnRNA. For a more detailed discussion of the experimental data concerning these conclusions see the papers by DARNELL et al. (1971b, 1973). Evidence for the presence of high molecular weight precursors to mRNA has come from experiments with cells transformed by SV40 (LINDBERG and DARNELL, 1970; TONEGAWA et al., 1970), where the virus genome apparently is covalently linked to host cell DNA (reviewed by GREEN, 1970). These cells produce high molecular weight HnRNA

Table 4. *Adenovirus Specific Nuclear RNA Late in Virus Infection*

RNA[a]	Total cpm × 10^6	Per cent of total	Fraction virusspecific
Not polyadenylated[b]	1.68	62	0.19
Polyadenylated	1.03	38	0.78

[a] Infected cells were labeled from 14—16 hours after infection with ^{3}H-uridine and nuclear RNA extracted according to HOLMES and BONNER (1973). The RNA was selected on oligo (dT) cellulose (AVIV and LEDER, 1972) and the fraction of adenovirus RNA was determined by exhaustive hybridization (LINDBERG et al., 1972).
[b] This refers to the fraction not retained by oligo (dT) cellulose of which ribosomal precursor RNA was estimated to constitute at least 60 per cent.

molecules with virus-specific sequences, while at the same time much smaller virus-specific mRNAs are found in protein synthesizing polysomes.

During the last years, further evidence suggesting the processing of HnRNA into mRNA has been collected. Polyribosomal mRNA as well as HnRNA contain poly (A) at the 3'terminus (EDMONDS and CARAMELA, 1969; KATES, 1970; DARNELL et al., 1971a and b; EDMONDS et al., 1971; LEE et al., 1971; PHILIPSON et al., 1971; MENDECKI et al., 1972; MOLLOY et al., 1972). A substantial fraction of HnRNA molecules of all size classes is polyadenylated and a significant fraction of poly (A) synthesized in the nucleus is eventually transported to the cytoplasm as part of mRNA molecules (JELINEK et al., 1973). All together, this supports the idea of HnRNA being a precursor of mRNA. The addition of poly (A) to nuclear RNA appears to be a prerequisite for the appearance of mRNA on polyribosomes (DARNELL et al., 1971b). In all these respects the adenovirus system resembles the host cell.

In the animal cell the major part of the HnRNA seems to turn over in the nucleus, leaving 10—20 per cent to be transported to the cytoplasm as mRNA (SOIERO et al., 1968; PENMAN et al., 1968). Late in adenovirus infection the major part of the nuclear nonribosomal RNA appears to be virus-specific (Table 4), and of the viral nuclear RNA *sequences* at least 75—80 per cent are preserved as mRNA. Kinetic evidence suggests, however, that only 20 per cent of the amount of viral RNA synthesized in the nucleus is transported to the cytoplasm late in the infectious cycle (PHILIPSON et al., 1974). Taken together, these two findings suggest either that only one of five viral transcripts are processed into mRNA or that each transcript is functionally monocistronic and processed into a single mRNA species. In both cases it appears that the major controls are at a posttranscriptional level. The former alternative suggests that the high molecular weight viral RNA found in the nucleus represents polycistronic precursors. Eight out of ten precursors become degraded whereas two out of ten molecules are cleaved into a number of specific fragments, which become polyadenylated and transported as mature mRNA to the cytoplasm. In the latter case one transcript is made for each mRNA species and the major part of the transcript turns over and only a small mRNA region is preserved. The latter model appears to operate for HnRNA in uninfected cells, but it has not been excluded that some cellular HnRNA molecules are processed and utilized to the large extent required for the former alternative (DARNELL et al., 1973).

The polysomal mRNA molecules of eukaryotic cells appear to have proteins associated with them (SPIRIN and NEMER, 1965; PERRY and KELLEY, 1968; HENSHAW, 1968; KUMAR and LINDBERG, 1972; LEBLEU et al., 1971; BLOBEL, 1973; MOREL et al., 1971, 1973; LINDBERG and SUNDQUIST, 1974) and there is evidence suggesting that the HnRNA in the nucleus also exists in the form of ribonucleoprotein complexes (SAMARINA et al., 1968; PERRY and KELLEY, 1968; PEDERSON, 1974). The possible association of proteins with adenovirus nuclear RNA has not yet been investigated, but polysomal mRNP particles from uninfected and adenovirus infected cells appear to have several major polypeptides in common. In addition to the common polypeptides, mRNP from infected cells harvested late in the infectious cycle appears to have one extra, possibly virus-specific polypeptide (LINDBERG and SUNDQUIST, 1974). The absolute specificity for the mRNA and the functional role of these polypeptides remain to be established.

In the animal cell 30—40 per cent of the total poly (A) containing RNA in the cytoplasm is *not* present in polyribosomes but instead in ribonucleoprotein particles sedimenting slower than polysomes (LINDBERG and PERSSON, 1972; JELINEK et al., 1973). It has been suggested that this non-polysomal poly (A) containing RNA in fact is mRNA on its way to be engaged in translation (SCHOCHETMAN and PERRY, 1972), but it has been difficult to obtain proof for this assumption. In adenovirus-infected cells in the late phase 40—70 per cent of the cytoplasmic virus-specific RNA is found in the non-polysomal pool (RASKAS and OKUBO, 1971; PHILIPSON et al., 1973). The amount of viral RNA in this region depends on the supply of amino acids in the medium. The richer the supply of amino acids the more of the mRNA is engaged in protein synthesis. The non-polyribosomal pool has been shown to contain the same pattern of major viral RNAs as the polysomes (PHILIPSON et al., 1973) (Fig. 13), and kinetic experiments suggest that the labeled viral RNA molecules appear more rapidly in the non-polyribosomal pools than in the polyribosomal fraction (PHILIPSON et al., 1973). These data together suggest that the non-polysomal pool of RNP particles contains mRNA *en route* to translation.

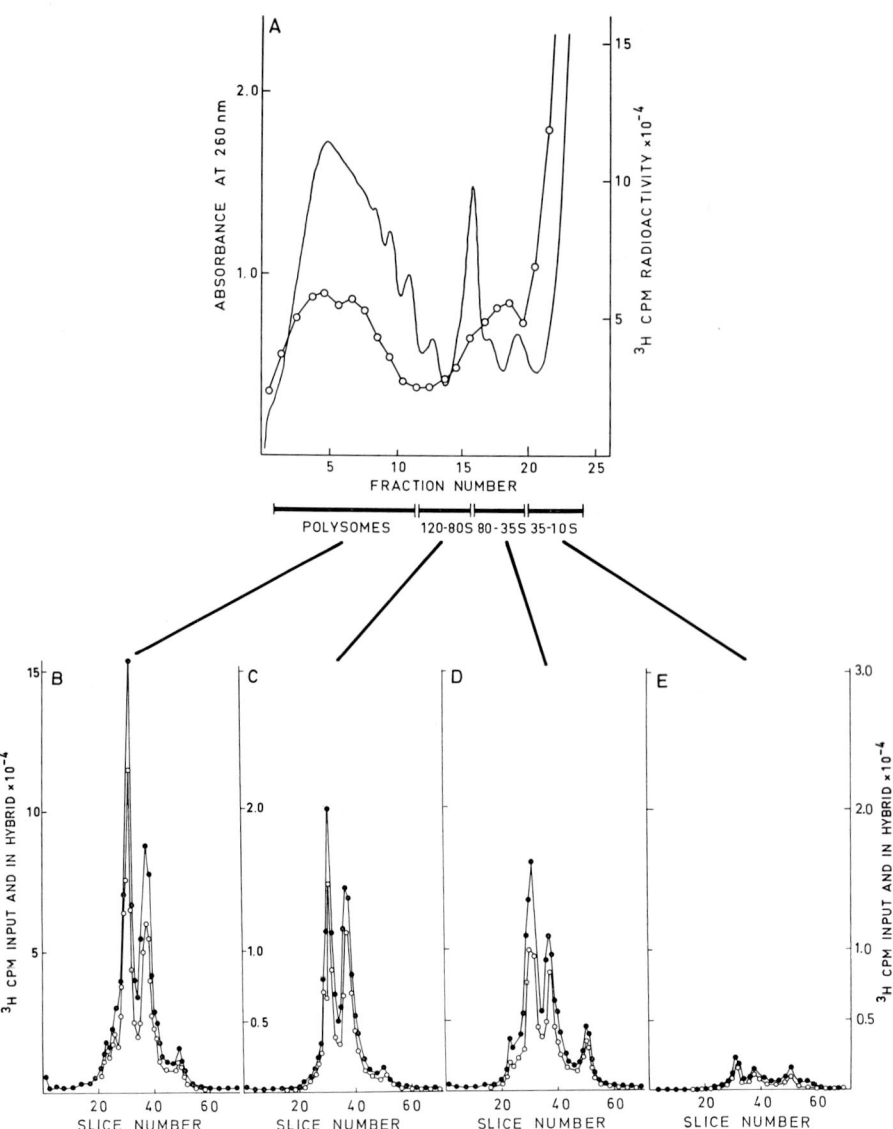

Fig. 13. Analysis of late viral RNA from polyribosomes and late viral cytoplasmic RNA not associated with polyribosomes.
Cells were labeled with ^{3}H-uridine from 14—16 hours after adenovirus infection and the cytoplasm was fractionated on a 7—47 per cent sucrose gradient (LINDBERG and PERSSON, 1972). The top panel shows the optical density (———) and the acid precipitable radioactivity (o———o). RNA from four indicated regions was deproteinized and the poly (A) containing RNA isolated by poly (U) Sepharose chromatography. After ethanol precipitation the RNA was analyzed on polyacrylamide gels and the electropherograms are shown in the lower frames.
•———• Total radioactivity; o———o Radioactivity in viral RNA

F. Adenovirus DNA Replication

During one step growth conditions viral DNA synthesis begins in the nucleus 6—8 hours after infection. With ad2 and ad5 the maximum rate of viral DNA synthesis is reached about 13 hours after infection, and at this time host cell DNA synthesis is suppressed so that 90 per cent or more of newly synthesized DNA is viral (GINSBERG et al., 1967; PIÑA and GREEN, 1969). No information is available about the enzymes involved in synthesis of adenovirus DNA and no virus-specific DNA polymerase has so far been identified. VAN DER VLIET and LEVINE (1973) have searched for DNA-binding proteins in adenovirus infected cells. Two polypeptides with molecular weights of 72,000 and 48,000 have been identified. Both have a strong preference for single stranded DNA and may thus be associated with the exposed single strands of adenovirus DNA during replication (see below). The temperature-sensitive mutant ts125 (section VIII) of ENSINGER and GINSBERG (1972) which fails to synthesize DNA at the nonpermissive temperature does not make these proteins. There is a difference between host cell and viral DNA synthesis with regard to the requirement for protein synthesis; HORWITZ et al. (1973) demonstrated that after viral DNA synthesis has begun it is no longer sensitive to cycloheximide, whereas continued host cell DNA synthesis requires simultaneous protein synthesis.

Most studies on adenovirus DNA synthesis have been carried out with the group C viruses mainly because the buoyant densities of their DNAs allow separation from host cell DNA by equilibrium centrifugation in CsCl (PIÑA and GREEN, 1969) or by chromatography on methylated albumin-kieselguhr (GINSBERG et al., 1967).

1. Characteristics of Adenovirus DNA during *in vivo* Replication

Replicating DNA has been isolated from nuclei of infected cells usually 12 to 13 hours after infection when the maximum rate of viral DNA synthesis is observed. DNA which is pulse-labeled with radioactive thymidine has been analyzed by equilibrium centrifugation in CsCl. Several investigators have noticed that short pulses of thymidine are incorporated into DNA which has a 5—10 mg/ml higher buoyant density than mature viral DNA (PEARSON and HANAWALT, 1971; SUSSENBACH et al., 1972; VAN DER EB, 1973; PETTERSSON, 1973). After long pulses or after short pulses followed by a chase with unlabeled thymidine the radioactivity is incorporated in structures which have the same density as mature viral DNA. Thus, intermediates in adenovirus DNA replication appear to have a greater buoyant density than mature DNA. The density difference can be eliminated after digestion with nucleases which are specific for single stranded nucleic acids but not after digestion with RNase (PETTERSSON, 1973) and it has been estimated that replicating intermediates contain 20—30 per cent single stranded DNA (PETTERSSON, 1973). Replicating DNA from CELO virus has been found to have similar properties (BELLETT and YOUNGHUSBAND, 1972). Because of their single stranded character, replicating intermediates can be selected by chromatography on benzoylated-naphtoylated-DEAE (BND)-cellulose (SUSSENBACH et al., 1972; BELLETT and YOUNGHUSBAND, 1972; VAN DER EB, 1973). In neutral sucrose gradients, replicating intermediates sediment faster than mature DNA (35—80S) whereas under denaturing conditions replicating DNA seems to consist of unit length and shorter DNA (HORWITZ, 1971; VAN DER EB, 1973; BELLETT and YOUNGHUSBAND, 1972). HORWITZ (1971) analyzed replicating DNA by alkaline sucrose gradient centrifugation after pulse-labeling for very short time periods. No distinct class of short fragments could be detected and thus no evidence could be obtained for discontinuous replication of adenovirus DNA. On the other hand BELLETT and YOUNGHUSBAND (1972) analyzed replicating CELO DNA under similar conditions and observed a peak of DNA fragments sedimenting at about 12S after labeling for short times. Pulse-chase experiments indicate that these fragments are intermediates in replication. It is conceivable that "Okazaki-like" fragments (OKAZAKI et al., 1968) of adenovirus DNA are shortlived and therefore difficult to detect, because synthesis of ad 2 DNA is more rapid than that of CELO virus DNA (BELLETT and YOUNGHUSBAND, 1972). In fact, SUSSENBACH and coworkers (personal communication) have detected "Okazaki-like" fragments after pulse-labeling of infected cells which had been arrested with hydroxyurea. WINNACKER (1975) has also shown that ad 2 DNA replicates with "Okazaki-like" fragments as intermediates in isolated nuclei.

VAN DER EB (1973) analyzed replicating DNA from adenovirus infected cells in the electron microscope. Two types of intermediates were observed: (i) Y-shaped molecules with an arm consisting of single stranded or partially single stranded DNA; (ii) Linear molecules of unit length which were either single stranded or consisted of duplex DNA with single stranded gaps. These observations are compatible with the model for adenovirus DNA replication which has been proposed by SUSSENBACH et al. (1972) and which is discussed below (Fig. 14).

Several investigators have looked for circular intermediates in adenovirus DNA replication by performing dye-buoyant density gradient centrifugation (DOERFLER et al., 1973; BELLETT and YOUNGHUSBAND, 1972). However, no covalently closed circular intermediates have so far been detected by this method or by electron microscopy (BELLETT and YOUNGHUSBAND, 1972). On the other hand, ROBINSON et al. (1973) have reported that a circular form of viral DNA can be isolated from virions. A protein appears to link the termini of the DNA and a similar protein may be involved in viral DNA replication to create circular intermediates. Such structures would, however, be difficult to detect since they would be linearized after deproteination of the DNA.

By following the fate of parental DNA or by continuous labeling *in vivo*, BURLINGHAM and DOERFLER (1971) demonstrated three different classes of intracellular viral DNA from ad 2 and ad 12 by neutral surcrose gradient centrifugation; one larger than viral DNA, viral DNA, and a third class smaller than viral DNA. The fast sedimenting component may represent viral DNA integrated in cellular DNA or DNA undergoing transcription (DOERFLER et al., 1973). A complex of viral RNA and DNA was isolated by CsCl-propidium iodide gradient centrifugation (DOERFLER et al., 1973) in which the RNA was of a size similar to mRNA and susceptible to RNase. The relationship of this complex to the rapidly sedimenting parental DNA is, however, unclear. BURGER et al. (1974) have extended these studies and provided evidence that the fast sedimenting component may represent adenovirus DNA integrated in the host chromosome. The slowly sedimenting population of DNA has an average sedimentation coefficient of 12S. The studies of BURLINGHAM and DOERFLER (1972) infer that the small viral DNA could be degradation products of the virion-associated endonuclease. SUSSENBACH (1971), on the other hand, has reported that fragment molecules of ad 12 DNA may replicate independently.

2. Replication of Adenovirus DNA in Isolated Nuclei

Isolated nuclei have the advantage, compared to intact cells, in that they are able to incorporate externally supplied nucleoside triphosphates into nucleic acids (FRIEDMANN and MUELLER, 1968; WINNACKER et al., 1971). In addition, pulse-chase experiments are facilitated in these systems and they allow the use

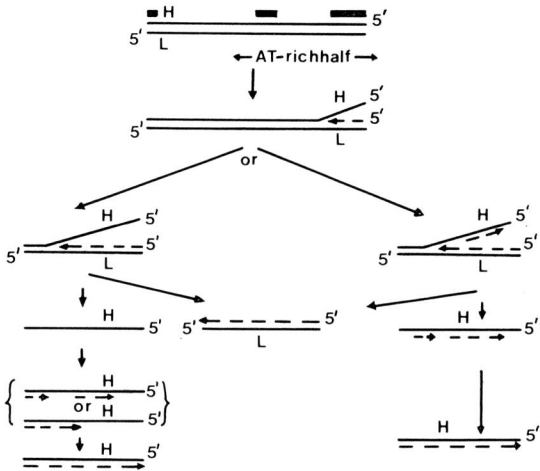

Fig. 14. Model for replication of ad 5 DNA.
Parental strands are drawn as solid lines and newly synthesized DNA is indicated with dashed lines. The black bars in the upper panel represent the three AT-rich areas as defined by the electron microscopical denaturation map (ELLENS et al., 1974). The complementary strand with low and high buoyant density in alkaline CsCl are designated as L and H (SUSSENBACH et al., 1973)[1].
DNA replication starts at the right hand end and the newly synthesized DNA displaces the H-strand. After prolonged displacement synthesis two types of intermediates arise, one containing a single stranded DNA arm on which complementary strand synthesis has started and another with a single stranded arm where no complementary strand synthesis has been initiated. After completion of the displacement synthesis both intermediates will yield one mature double stranded molecule and one new intermediate which is either entirely single stranded or double stranded with single stranded gaps.
The latter intermediate are completed by discontinuous replication. Recent evidence suggests that the displacement synthesis on the L-strand may also be discontinuous (SUSSENBACH, personal communication.) Redrawn from ELLENS et al. (1974)

[1] The L-strand has a higher buoyant density than the H-strand in neutral CsCl in the presence of ribopoly (U, G). Thus, the L-strand corresponds to the h-strand as defined by TIBBETTS et al. (1974) and as indicated in Figs. 11 and 12. Thus, the maps which are illustrated in Figs. 11 and 12 are oriented in the same way as the parental DNA in this Figure.

of inhibitors of macromolecular synthesis, which do not penetrate the plasma membrane. In these nuclear preparations already initiated adenovirus DNA molecules appear to be elongated at a slow rate but initiation of new rounds of DNA synthesis does not seem to occur (VAN DER VLIET and SUSSENBACH, 1972).

An extensive characterization of replicating DNA in isolated nuclei from cells infected with ad5 has been carried out by SUSSENBACH and coworkers (SUSSENBACH and VAN DER VLIET, 1972; VAN DER VLIET and SUSSENBACH, 1972; SUSSENBACH et al., 1972; SUSSENBACH et al., 1973; ELLENS et al., 1973; SUSSENBACH and VAN DER VLIET, 1973). The replication of adenovirus DNA in the isolated nuclei is semiconservative (VAN DER VLIET and SUSSENBACH, 1972) as has also been shown to be the case *in vivo* for CELO virus DNA (BELLETT and YOUNGHUSBAND, 1972). Replicating DNA has been selected by BND-cellulose chromatography and characterized by electron microscopy. Three classes of replicating molecules resembling those described by VAN DER EB (1973) have been observed (SUSSENBACH et al., 1972): The *first* class consists of linear Y-shaped molecules with an arm of single stranded or partially single stranded DNA. The *second* class consisted of unbranched linear molecules, partly double stranded and partly single stranded, with varying ratios of double stranded to single stranded regions. The *third* class also contained unbranched linear molecules, some of which were purely double stranded and some purely single stranded. On the basis of these studies, SUSSENBACH et al. (1972) have suggested a model for the replication of adenovirus DNA (Fig. 14). Replication starts from one end, and one of the parental strands is displaced by the growing strand giving rise to Y-shaped structures. At a later stage, synthesis is initiated on the displaced strand which becomes detached from the replicating structure. The displaced strand is converted to duplex DNA by discontinuous synthesis (Fig. 14). This model is compatible with data obtained *in vivo* for CELO, ad2 and ad5 DNA replication (BELLETT and YOUNGHUSBAND, 1972; VAN DER EB, 1973; PETTERSSON, 1973). Recently, SUSSENBACH et al. (1973) have also presented evidence that replication starts at one specific end and only the strand which bands at a higher density in alkaline CsCl gradients seems to be displaced. In the ad2 system this strand corresponds to the l-strand as defined by buoyant density in CsCl after binding to poly (U, G) (TIBBETTS et al., 1974). Further evidence corroborating this notion has been obtained by electron microscopic investigations of partially denatured replicating molecules (SUSSENBACH et al., 1973). In this study, the end where replication starts (*i.e.* the end containing the 5' terminus of the displaced heavy strand) was found to correspond to the AT-rich end of ad5 DNA. This is the molecular right end as defined in the denaturation map of the ad5 DNA (ELLENS et al., 1974).

Although these data appear compelling, it is appropriate to stress that the DNA synthesis occurring in the isolated nuclei reflects only elongation of already initiated DNA molecules and that probably no reinitiation takes place. Only 10 per cent of the parental DNA, which is present in the nucleus, appears to be active in elongation. This means that there are some important factor(s) missing in this system and the structures observed could differ from those present *in vivo*. At present, no information is available with regard to the initiation of adenovirus DNA replication. If DNA synthesis is primed with RNA as seems to be the case in several other systems (BRUTLAG et al., 1972; SUGINO et al., 1972; WICKNER *et al.*, 1972; MAGNUSSON *et al.*, 1973), the problem arises how to complete the 5' end of the daughter strand after digestion of the primer. As pointed out by BELLETT and YOUNGHUSBAND (1972), it is difficult to conceive how a linear duplex DNA-like adenovirus DNA is able to prime repair synthesis at the 5' end of the daughter strand. There is no terminal redundancy in adenovirus DNA and concatemers like those existing during replication of bacteriophage T7 DNA cannot form. Thus, the completion of the newly synthesized strands cannot take place according to the model proposed for this phage (WATSON, 1972), which also replicates with linear intermediates. BELLETT and YOUNGHUSBAND (1972) have suggested that the completion of the 5' ends of nascent molecules involves formation of concatemers with staggered nicks in both strands at genome intervals. Possibly the protein which can circularize adenovirus DNA (ROBINSON *et al.*, 1973) may facilitate formation of the proposed concatemers.

G. Translation

The adenoviruses are assembled in the nucleus, but all the proteins appear to be synthesized in the cytoplasm (THOMAS and GREEN, 1966; VELICER and GINSBERG, 1968). Adenovirus specific mRNA and newly synthesized viral polypeptides are predominantly found in association with non-membrane bound polyribosomes (THOMAS and GREEN, 1966; VELICER and GINSBERG, 1968; HORWITZ *et al.*, 1969). Shortly after synthesis, the polypeptides are assembled into viral structural proteins in the cytoplasm and then rapidly transported to the nucleus for assembly of virions (VELICER and GINSBERG, 1970). Since some temperature sensitive mutants accumulate viral proteins in the cytoplasm, the transport may require a virus-coded function (ISHIBASHI, 1970; RUSSELL *et al.*, 1972a).

1. Early Proteins

The majority of the proteins synthesized early after infection are host proteins, which makes it difficult to identify early viral polypeptides. However, utilizing sera from hamsters with adenovirus-induced tumors, virus-specific antigens have been detected early after infection by complement fixation and immunofluorescence. The T-antigen, where T stands for tumor, is one of the few early viral products which so far has been identified (ROUSE and SCHLESINGER, 1967; RUSSELL *et al.*, 1967a; POPE and ROWE, 1964; SHIMOJO *et al.*, 1967). This antigen reacts with sera from tumor-bearing animals and seems to be present in adenovirus-induced tumors as well as early during lytic infection. In contrast to the papova virus system, where T-antigen has been detected only in the nucleus, the adenovirus T-antigen is found both in the nucleus and in the cytoplasm. T-antigen is synthesized also in the presence of inhibitors of DNA synthesis (GILEAD and GINSBERG, 1965; FELDMAN and RAPP, 1966) and is therefore to be considered as a true early product. GILEAD and GINSBERG (1968a and b) have purified adenovirus 12 T-antigen and described a single component which sediments at 2.6S. Other investigators have claimed that the adenovirus T-antigen consists of several polypeptides (TOCKSTEIN *et al.*, 1968). Another antigen, the

P-antigen, has also been detected early after infection (RUSSELL and KNIGHT, 1967). This antigen appears to constitute a mixture of T-antigen and one of the viral core proteins (RUSSELL and SKEHEL, 1972).

RUSSELL and SKEHEL (1972) have analyzed the virus-induced proteins in ^{35}S-methionine labeled extracts from infected cells by SDS-polyacrylamide gel electrophoresis. Five polypeptide bands could be detected which were different from the virion polypeptides. They were specific for infected cells and at least one of these polypeptides seems to be an early product. The early polypeptides have also been studied when host cell protein synthesis was suppressed by a concomitant poliovirus infection in the presence of guanidine (BABLANIAN and RUSSELL, 1974). A polypeptide with a molecular weight of 64,000 was found to be synthesized early in infection. This polypeptide appears to be the major component of the P-antigen. WALTER and MAIZEL (1974) have reported on a similar study; infected cells were labeled with ^{35}S-methionine and analyzed by high resolving SDS-polyacrylamide gels. These investigators were able to detect three early virus-induced polypeptides. These polypeptides which were designated E_1, E_2 and E_3 have molecular weights of 70,000, 19,000 and 10,000, respectively. Polypeptides E_1 and E_2 are preferentially found in the cytoplasm whereas E_3 appears to be a nuclear protein. Polypeptide E_2 has been found to be a glycoprotein (ISHIBASHI and MAIZEL, 1974b). Polypeptide E_1 is probably identical with the polypeptide detected by RUSSELL and SKEHEL (1972) and a similar polypeptide has been identified by ANDERSON et al. (1973).

In growth arrested human embryo kidney cells or monkey kidney cells, adenovirus infection causes an increase in the synthesis of at least two of the enzymes (thymidine kinase and DNA-polymerase) involved in DNA synthesis early after infection (TAKAHASHI et al., 1966; LEDINKO, 1967; KIT et al., 1967; BRESNICK and RAPP, 1968; OGINO and TAKAHASHI, 1969). The increase is concurrent to the increase in cell DNA synthesis observed in these cells after virus infection. However, a similar effect cannot be detected after adenovirus infection of cells grown in suspension cultures. VAN DER VLIET and LEVINE (1973) have isolated two early polypeptides with molecular weights of 48,000 and 72,000 which have strong affinity for single stranded DNA. These proteins are presumably involved in adenovirus DNA replication and were described in a previous section (V: F). No virus-specific enzyme has been identified early in infection, although there are genetic evidence for the presence of genes coding for catalytic functions. There are several adenovirus mutants which in mixed infection with wild type give rise to the normal wild type yields independent of large variations in the ratio between the multiplicities of infection of mutant and wild type (WILLIAMS, personal communication).

2. Late Proteins

The shift from early to late viral gene expression occurs at around 8—10 hours after a synchronized infection in suspension cultures of KB or HeLa cells. At this time the first capsid proteins can be detected and 2—3 hours later the first progeny virus appears (WILCOX and GINSBERG, 1963c; POLASA and GREEN, 1965; RUSSELL et al., 1967b; EVERITT et al., 1971; RUSSELL and SKEHEL, 1972). The synthesis of the late proteins requires concomitant viral DNA synthesis and no late proteins can be observed in the presence of inhibitors like cytosine arabinoside, hydroxyurea or 5-fluorodeoxyuridine (GREEN, 1962a; FLANAGAN and GINSBERG, 1962; WILCOX and GINSBERG, 1963c). The major capsid units, hexon, penton and fiber, are synthesized in large excess (for a review see PHILIPSON and PETTERSSON, 1973), while the synthesis of some of the smaller proteins including the core proteins appear to be less extensive (WHITE et al., 1969; EVERITT et al., 1971). Only 5—20 per cent of the viral capsid proteins synthesized during the infectious cycle become incorporated into mature virions (GREEN, 1962b; WHITE et al., 1969; HORWITZ et al., 1969). After synthesis of the polypeptides, the viral proteins are rapidly assembled into structural units, which within a few minutes after completion are transported to the nucleus (VELICER and GINSBERG, 1970; HORWITZ et al., 1969).

Earlier results gave no indications that the structural proteins were produced by cleavage of larger precursors (HORWITZ et al., 1969; WHITE et al., 1969; RUSSELL and SKEHEL, 1972). Recent results, however, suggest that minor cleavage of precursor polypeptides occurs shortly after synthesis or during assembly (ANDERSON et al., 1973). The polypeptide VII, which is the major arginine-rich core protein (molecular weight 18,000) (LAVER, 1970; PRAGE and PETTERSSON, 1971), appears to be generated from a 21,000 molecular weight precursor (PVII in Fig. 2). One of five methionine containing tryptic peptides in the precursor is missing in the mature polypeptide VII (ANDERSON et al., 1973). The hexon-associated polypeptides VI (molecular weight 23,000) and VIII (molecular weight 13,000) have been proposed to be generated in a similar way (Fig. 2) (ANDERSON et al., 1973; EVERITT and PHILIPSON, 1974; ISHIBASHI and MAIZEL, 1974a; ÖBERG et al., 1975).

In addition to the polypeptides, which form structural units of the virion, at least 5 but possibly as many as 7 polypeptides can be detected late after infection in ^{35}S-methionine labeled cytoplasmic extracts (RUSSELL and SKEHEL, 1972; ANDERSON et al., 1973). These are absent or present in lower amounts in uninfected cells. One of the induced polypeptides is identical in size to the 100,000 molecular weight polypeptide which is found in messenger ribonucleoprotein particles in infected cells (LINDBERG and SUNDQUIST, 1974). The others have been claimed to be precursors to other polypeptides which are present in mature virions (WALTER and MAIZEL, 1974; ISHIBASHI and MAIZEL, 1974a).

Late in infection large amounts of viral proteins accumulate in the nucleus, where assembly of the virions takes place. As a reflection of this large crystalline structures, so called paracrystals, appear in the nucleus of the infected cell (KJELLÉN et al., 1955; MORGAN et al., 1957). The paracrystalline structures may be of different kinds and there is no general agreement on their nature. Crystalline areas of partially assembled virus (BOULANGER et al., 1973) and large crystalline areas of structural proteins, probably hexons or pentons (BOULANGER et al., 1970), have been observed. Crystalline structures have also been described which accumulate after the peak of viral synthesis in the productive cycle and may contain the major core protein (MARUSYK et al., 1972). WILLS et al. (1973) recently showed that the paracrystals in ad5 did not appear at the non-permissive temperature in cells infected with a *ts* mutant defective in fiber synthesis.

In addition to complete particles and structural proteins several other aggregated virus structures may be found in extracts of infected cells. Pentons, which have assembled into star-like structures with 12 pentons, so called dodecons, have been recovered from cells after infection with subgroup B and D adenoviruses (NORRBY, 1966, 1969b; LIEM et al., 1971). Empty capsids and incomplete particles, which contain no DNA or reduced amounts of viral DNA, are synthesized preferentially in cells infected with group A and B viruses (MAK, 1971; PRAGE et al., 1972; WADELL et al., 1973). The empty capsids lack the two core proteins, polypeptides V and VII (MAIZEL et al., 1968b; PRAGE et al., 1972; SUNDQUIST et al., 1973b; WADELL et al., 1973), and contain polypeptides (Fig. 2) which have been proposed to be precursors of polypeptides which are present in complete virions (SUNDQUIST et al., 1973b). The empty capsids are further discussed in section V:I.

3. Modification of Viral Polypeptides

There is evidence suggesting that the fiber polypeptide is both phosphorylated (RUSSELL et al., 1972b) and glucosylated (ISHIBASHI and MAIZEL, 1974b), and that the modification of the polypeptide probably occurs subsequent to translation. Only N-acetyl-glucosamine has been found associated with the fiber and there appears to be two residues of sugar per fiber polypeptide. One early virus specific non-structural polypeptide (E_2) has also been found to be labeled with ^{3}H-glucosamine (ISHIBASHI and MAIZEL, 1974b). Two or three other virus-induced non-structural polypeptides appear to be phosphorylated and the polypeptide of molecular weight 64,000 (RUSSELL and SKEHEL, 1972) is the major phosphorylated component (RUSSELL et al., 1972b) late in the infectious cycle. This polypeptide may correspond to the 71 K polypeptide band of ANDERSON et al. (1973, see Fig. 2).

4. Translation of Adenovirus mRNA in vitro

In vitro synthesis of adenovirus polypeptides has been studied in crude extracts from adenovirus infected cells which contain polyribosomes (CAFFIER and GREEN, 1971; CAFFIER et al., 1971; OKUBO and RASKAS, 1972; WILHELM and GINSBERG, 1972). Polypeptide synthesis appears to be initiated with methionine tRNA and most of the polypeptides of the virion can be detected after *in vitro* synthesis. WILHELM and GINSBERG (1972) also reported that capsomere assembly occurs in such systems; hexon and fiber polypeptides were assembled into mature proteins but no assembly of pentons could be detected. OKUBO and RASKAS (1972) used a reconstituted *in vitro* system from adenovirus infected cells still dependent on endogeneous messenger RNA. Supernatant fractions from infected and uninfected cells were analyzed in mixtures with the polyribosome preparations from infected cells. No difference in translation could be detected when enzyme fractions from infected and uninfected cells were exchanged.

In order to study control mechanisms operating in translation, it is desirable to have *in vitro* systems which depend on exogeneous messenger RNA. Recently it has been possible to translate isolated messenger RNA from adenovirus infected cells into structural proteins using *in vitro* systems which are dependent on external messenger RNA. ERON et al. (1974a) demonstrated that late mRNA selected on oligo (dT) cellulose can direct the synthesis of hexon, penton base, IIIa and fiber polypeptides in an *in vitro* system derived from ascites cells. The

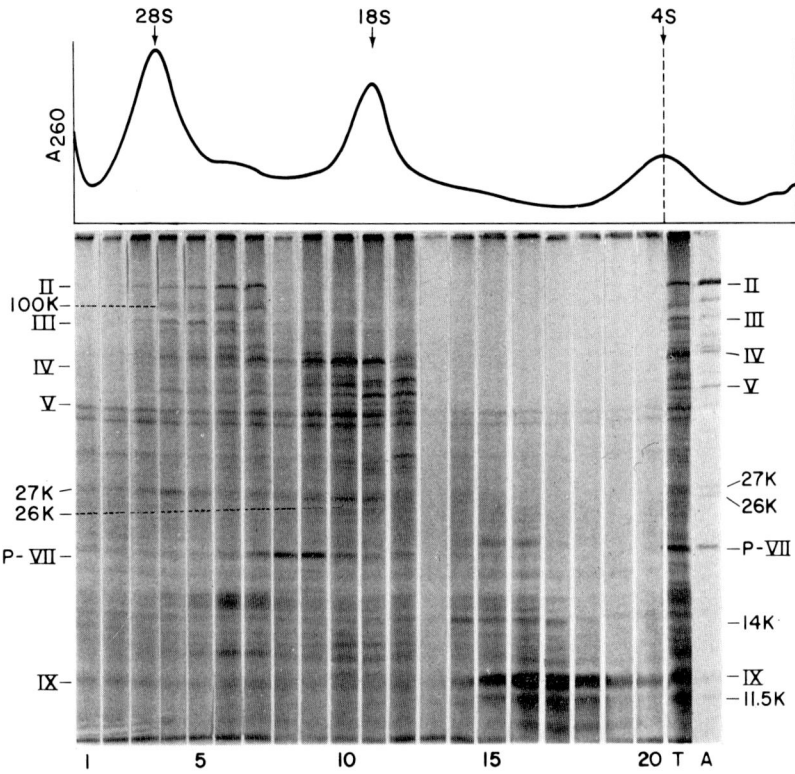

Fig. 15. SDS-polyacrylamide gel autoradiogram of the *in vitro* synthesized polypeptides which were programmed by fractionated ad 2 mRNA.
RNA was fractionated on formamide-containing sucrose gradients (ANDERSON et al., 1974). RNA was melted in 90 per cent formamide and layered on a 5—20 per cent linear sucrose gradient in 50 per cent formamide. Each fraction was precipitated with ethanol and dissolved in water. A mammalian cell-free system was programmed with RNA from each fraction and the products were analyzed on SDS polyacrylamide gels. The gel pattern is aligned with the optical density profile of the sucrose gradient. Fraction numbers are given at the bottom of the Figure. Polypeptides programmed by unfractionated cytoplasmic ad 2 mRNA (column T) and an *in vivo* labeled sample of polypeptides from ad 2-infected HeLa cells (column A) are shown for comparison.
Reprinted from ANDERSON et al. (1974)

products which range in molecular weights from 20,000 to 120,000 daltons were shown to be identical with the polypeptides of the structural proteins by SDS gel electrophoresis and by mapping of tryptic peptides. Furthermore, the *in vitro* synthesized polypeptides had the same antigenic specificity as the structural proteins and could be precipitated with specific antisera (ERON et al., 1974a and b;

ÖBERG et al., 1975). In a separate study it was found that core polypeptide V as well as the precursor to the major core protein, P VII (ANDERSON et al., 1973) (see Fig. 2) also were synthesized in vitro (ERON et al., 1974b). ERON and WESTPHAL (1974) also developed a reversible hybridization technique by which viral specific RNA could be selected. Such RNA directs the synthesis of several capsid proteins demonstrating that they are specified by the viral genome.

ANDERSON et al. (1974) demonstrated with a purified in vitro system, similar to that described by SCHREIER and STAEHELIN (1973), that late viral RNA directs the synthesis of all the major structural polypeptides with a very low background. The major class of late viral mRNA with a sedimentation constant of around 26S may code for several polypeptides, the hexon (II) and penton base (III) as well as the virus-induced 100K and 27K polypeptides (Fig. 2). The 19—22S area seemed to code preferentially for the polypeptides IV (fiber), V, 26K and the precursor to polypeptide VII. The 9S RNA almost exclusively codes for polypeptide IX during in vitro synthesis (Fig. 15) (ANDERSON et al., 1974; ÖBERG et al., 1975). The hexon polypeptide is less efficiently synthesized in the wheat germ system (ANDERSON et al., 1974; ÖBERG et al., 1975), than in the ascites cell (ERON et al., 1974a) and the purified system (ANDERSON et al., 1974). The components in the cell-free system must work with considerable accuracy to achieve faithful synthesis of the hexon polypeptide which contains 900—1000 peptide bonds (MAIZEL et al., 1968a; JÖRNVALL et al., 1974b).

In vitro systems for protein synthesis will hopefully be of help to identify products which are coded by individual RNAs. These systems will also provide tools to study control mechanisms operating in translation of viral messenger RNA. It has been proposed that such mechanisms exist in vivo with regard to the preferential synthesis of viral over cell proteins and also with regard to the preferential synthesis of hexon polypeptides over other virion polypeptides (PERLMAN et al., 1972) as will be discussed in section V:I:2. The fact that the large 26S mRNA class is deficient in its association with polyribosomes in translation defective monkey cells infected with ad2 (HASHIMOTO et al., 1973) suggests a specific control for initiation of hexon translation (see section V:I:2).

H. Host Cell Macromolecular Synthesis

Both primary cells in monolayer and established cell lines in suspension fail to divide after infection with adenoviruses. This effect can probably be ascribed to the generalized interference with the synthesis of host cell macromolecules in the late phase of infection. In growth arrested monolayer cultures adenovirus infection, like SV40 and polyoma virus infection, results in early stimulation of cell DNA synthesis as well as of the proteins involved in DNA replication (LEDINKO, 1967, 1970). No such effect is seen after infection of cells grown in suspension cultures. Synthesis of ribosomal RNA is increased during the early phase of the infection in primary cells (LEDINKO, 1972), and possibly also in suspension cultures (RASKAS et al., 1970). During the late phase, on the other hand, the synthesis of host cell macromolecules is drastically inhibited.

At 6—8 hours after infection with adenovirus, the synthesis of cellular DNA begins to decline and by 10—13 hours post infection about 90 per cent of the newly synthesized DNA is viral (GINSBERG et al., 1967; PIÑA and GREEN, 1969). HODGE and SCHARFF (1969) have investigated the effect of adenovirus infection on host cell DNA synthesis in synchronized HeLa cells. When viral DNA synthesis occurs during the G1 phase of the growth cycle, cellular DNA replication is not induced in the subsequent S phase or later. When viral DNA synthesis is timed to begin during the S phase, the cellular DNA synthesis starts but the round of cellular DNA replication fails to go to completion. These results imply that already initiated cell DNA synthesis can continue, but that the initiation of cellular DNA synthesis is inhibited.

Host cell protein synthesis declines at about the same time as cell DNA synthesis (GINSBERG et al., 1967). Pulse-labeling experiments with radioactive amino acids have shown that the label almost exclusively enters virus-coded or induced proteins already 12 hours after infection (RUSSELL and SKEHEL, 1972; ANDERSON et al., 1973). Some host cell mRNA appears to be translated also late in the infectious cycle. For instance, the protein moiety of the messenger ribonucleoprotein particles (mRNP) is synthesized late in infection at the same rate as in uninfected cells (LINDBERG and SUNDQUIST, 1974).

Concurrent with the reduction in host protein synthesis, the accumulation of newly synthesized ribosomal RNA in the cytoplasm is suppressed. At around 12 hours after infection only 10—12 per cent of the normal amount of ribosomal RNA is transported to the cytoplasm (RASKAS et al., 1970; PHILIPSON et al., 1973). Since a significant amount of the 45S precursor to ribosomal RNA is still synthesized (RASKAS et al., 1970; LEDINKO, 1972), the interference with ribosome synthesis appears to effect primarily the cleavage of the ribosomal precursor RNA, but a suppression of synthesis of the 45S precursor has also been observed (ELICERI, 1973). It is not clear whether the interference with the processing of ribosomal RNA is a result of the inhibition of host cell protein synthesis or whether there is some specific inhibitory mechanism. It is, however, known that conditions which suppress host cell protein synthesis in uninfected cells also effects the maturation of ribosomal RNA (MADEN et al., 1969; WILLEMS et al., 1969). Synthesis of cellular transfer RNA (tRNA) occurs unabatedly throughout the infectious cycle (GINSBERG et al., 1967; MAK and GREEN, 1968) and no virus-specific tRNA has been recognized (RASKA et al., 1970; KLINE et al., 1972).

It has been suggested that the structural proteins of the virus are involved in the inhibition of host macromolecular synthesis. LEVINE and GINSBERG (1967, 1968) reported that the fiber protein inhibits RNA and DNA polymerases *in vitro*. It was also found that high concentrations of purified fiber could suppress the macromolecular synthesis in uninfected cells. However, very high concentrations were required and the inhibition of host cell macromolecular synthesis during infection with adenoviruses usually starts before large amounts of fiber protein have accumulated. Furthermore, ts-mutants which have a defect in synthesis of fiber protein or viral DNA still inhibit host cell DNA synthesis at non-permissive temperatures (WILKIE et al., 1973).

I. Control Mechanisms

As discussed in a previous section, no information is yet available on the mechanisms controlling the switch from early to late phase during productive infection. The key event for the transition to the late phase is the triggering of viral DNA replication. If this is prevented, there is no switch from early to late expression of the viral genome. Concomitant to the increase in viral DNA synthesis, host cell DNA synthesis starts to decline. The appearance of increasing amounts of late messenger RNA coincides in time with a decreased production of mature ribosomes and host mRNA. In all these steps specific control mechanisms are implied. In addition, there is evidence for regulation at the level of translation.

1. RNA Synthesis and Processing

As outlined above, the adenovirus genome appears to be transcribed into large viral RNA molecules found in the nucleus. These mRNA precursors may be polycistronic and cleaved at specific points along the polynucleotide chains to give rise to the mature mRNA. Alternatively, only a single mRNA molecule is generated from each transcript and most of the RNA becomes degraded.

More than 90 per cent of the RNA, which is synthesized and transported to the cytoplasm late in the infection, is adenovirus-specific (LINDBERG et al., 1972; PHILIPSON et al., 1973) which suggests a preferential transcription of viral DNA or preferential processing and transport of viral transcripts. In the nucleus only 10—20 per cent of the newly synthesized non-ribosomal RNA has been claimed to be virus-specific (PRICE and PENMAN, 1972a; McGUIRE et al., 1972), which would infer a virus-controlled event leading to preferential selection of viral mRNA for transport to the cytoplasm. Recent results in our laboratory, however, suggest that a larger fraction of nuclear RNA is of viral origin (Table 4) emphasizing that a control mechanism also exists at the level of transcription (PHILIPSON et al., 1974).

2. Translation

At least two control mechanisms seem to be operating in translation of viral mRNA: (i) the switch during the late phase to preferential translation of viral mRNA, and (ii) preferential translation of individual viral polypeptides.

Late in infection, protein synthesis continues at an undiminished rate, but the polysomes become almost exclusively engaged in the synthesis of viral proteins. The transition in protein synthesis from the host to the virus pattern is a rapid process which only takes 2 hours at 8—10 hours post infection (RUSSELL and SKEHEL, 1972; ANDERSON et al., 1973). A large fraction of the host mRNA appears to turn over slowly in uninfected cells, on the average once per cell generation (GREENBERG, 1972; SINGER and PENMAN, 1972). A similar turnover rate has been found for cellular mRNA in infected cells (CRAIG et al., 1974; PHILIPSON et al., 1974). This finding suggests that the switch from early to late translation is not only due to preferential export of viral mRNA from the nucleus. An additional mechanism ensuring the preferential utilization of the viral mRNA in the cytoplasm has to be in operation. Whether this means that there are virus-specific initiation factors for translation or that there is an inhibition of elongation of host mRNA is not clear. The observation that vaccinia virus transcription is unaffected but translation of vaccinia virus mRNA is deficient in mixed infections with vaccinia and adenovirus suggests a preferential initiation of adenovirus mRNA (GIORNO and KATES, 1971). The understanding of the control mechanisms discussed here requires an *in vitro* protein synthesizing system, where initiation factors from virus infected and uninfected cells can be compared.

It has been suggested from the work of several investigators including PERLMAN *et al.* (1972) that the hexon polypeptide is preferentially synthesized at 37° C at a rate 2—4 times faster than that of the fiber polypeptide (EVERITT *et al.*, 1971). This could reflect the presence of different amounts of mRNA for different polypeptides. However, in the presence of low concentrations of cycloheximide the relative rate of synthesis of the different viral polypeptides is altered compared with controls without the drug. Since cycloheximide slows elongation without affecting initiation, this would mean that under normal conditions there are differences in the rate of initiation of the different polypeptides (PERLMAN *et al.*, 1972).

Some mRNA molecules are much larger than one would expect from the size of the polypeptides they code for. The largest adenovirus polypeptide is that of the hexon which has a molecular weight of about 100,000—120,000 (HORWITZ *et al.*, 1970; JÖRNVALL *et al.*, 1974b). The largest viral mRNA has a molecular weight of about 2×10^6 (LINDBERG *et al.*, 1972), which theoretically could code for a polypeptide of molecular weight 200,000. The predominant size class of late mRNA has a molecular weight around 1.5×10^6 (26S) which is larger than would be required to synthesize any of the viral polypeptides. Thus, some of the viral mRNAs may be polycistronic. This is supported by the work of ANDERSON *et al.* (1974) who analyzed viral mRNA which had been fractionated on sucrose gradients in an *in vitro* protein synthesizing system. In some cases the mRNA was found to be twice as long as would be required to code for the polypeptides synthesized (Fig. 15). In picornavirus infected cells the viral mRNA is translated into one large protein precursor molecule, which through cleavage is processed to mature viral proteins (SUMMERS and MAIZEL, 1968; HOLLAND and KIEHN, 1968; JACOBSON and BALTIMORE, 1968). There is so far no evidence in the adenovirus system for the presence of large precursor proteins of the type seen in picornavirus infected cells. However, minor proteolytic trimming occurs with at least the precursors to the virion polypeptides VI, VII and VIII (ANDERSON *et al.*, 1973; ISHIBASHI and MAIZEL, 1974a; EVERITT and PHILIPSON, 1974) as was discussed in a previous section (V:G:2).

J. Assembly of Adenoviruses

1. Mechanism of Assembly

Late during adenovirus infection, both complete virus particles and capsids which lack viral DNA (empty capsids) accumulate in the nucleus. The proportion of empty capsids formed varies among different serotypes. After infection with ad 2 only minute amounts are found, whereas a large pool of empty capsids is found after infection with ad 3 and ad 16 (PRAGE et al., 1972; WADELL et al., 1973). HORWITZ et al. (1969) reported that newly synthesized core polypeptides appear more rapidly in mature virus particles than the polypeptides of the capsid proteins. In agreement with this, results have been obtained (SUNDQUIST et al., 1973b; ISHIBASHI and MAIZEL, 1974a), which suggest that the empty capsids are intermediates in the assembly of virions, and that the DNA core probably enters preformed capsids. This conclusion was based on several lines of evidence: 1) The increase in empty capsids coincides in time with a rapid increase in the amount of virions. This indicates that the empty capsids are not created by breakdown of virions late in the infectious cycle. 2) During continuous labeling with radioactive amino acids, radioactivity appears first in empty capsids and after a lag period in complete virions. 3) Pulse-chase experiments with labeled and unlabeled amino acids showed that radioactivity increased first in empty capsids and could be chased into mature virions. The lag period before the appearance of label in virions was about 1 hour. All these results thus provide evidence for a precursor-product relationship between the empty capsids and the mature virions, but it is difficult to exclude that the true intermediate is a fragile structure which releases its DNA during purification.

Additional evidence for empty capsid intermediates has been obtained through the analysis of the polypeptide composition of these particles. The empty capsids lack the core proteins V and VII and also polypeptides VI, VIII and X, but contain polypeptides not present in the matuie virion (SUNDQUIST et al., 1973 b) (Fig. 16). It was recently proposed that three of the virion components (polypeptides VI, VII and VIII) are derived by cleavage of precursor polypeptides (ANDERSON et al., 1973; EVERITT and PHILIPSON, 1974; ISHIBASHI and MAIZEL, 1974 a; ÖBERG et al., 1975). In comparing these results with those of SUNDQUIST et al. (1973 b) it seems that the polypeptides 27 K and 26 K (Fig. 2) which probably are precursors to polypeptide VI and VIII, respectively, are present in the empty capsids. Polypeptide cleavage seems to take place after capsid assembly and possibly also after the DNA has entered the capsid. ISHIBASHI and MAIZEL (1974a) have proposed that the empty capsid becomes a "young virion" after receiving its DNA. This particle has the same density as the mature virion but its precursor polypeptides are uncleaved. It is noteworthy that the ratio of hexon to fiber polypeptide in the empty capsids is lower than in the virions suggesting that some hexons in the facets of the icosahedron might be missing (SUNDQUIST et al., 1973b). This could reflect discontinuities in the capsid through which the DNA core is introduced. It has not been possible by electron microscopy to establish where these discontinuities are located. Ts-mutants of ad5, which do not assemble at the non-permissive temperature, have been described (RUSSELL et al., 1972a). The mutants of the serological class 1 make most of the virion proteins but still do not produce infective virions. These mutants may aid to clarify the assembly pathway (see section VIII).

The serotypes which give rise to large amounts of empty capsids, like subgroup B viruses ad 3 and ad 16, also accumulate a heterogeneous population of particles, which in CsCl have buoyant densities intermediate to those of empty capsids

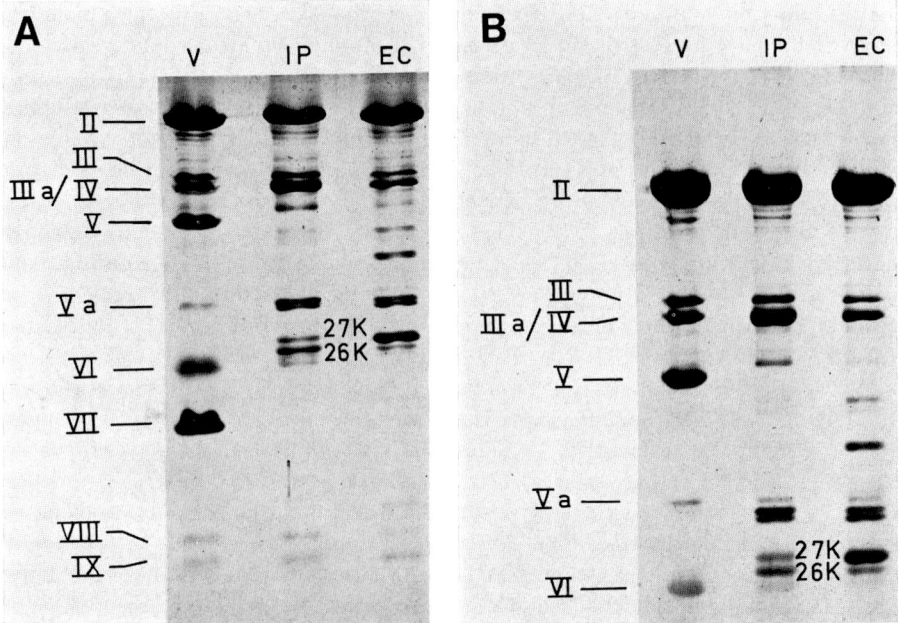

Fig. 16. Polypeptide composition of empty, incomplete and complete particles of ad 3. Cells infected with adenovirus type 3 contain in addition to *virions*, large amounts of particles without DNA, *empty capsids* and also particles with less DNA than mature virions *incomplete particles*.
Purified preparations of each class were analyzed by SDS-polyacrylamide gel electrophoresis. Frame A shows 13 per cent and frame B 10 per cent polyacrylamide gels. Ad 3 virions (V) shows a similar pattern as ad 2 virions (see Fig. 2). The empty (EC) and the incomplete (IP) particles lack the two core proteins V and VII and contain less polypeptide VI. The precursor polypeptides 27 K and 26 K which generate polypeptides VI and VIII, respectively (ANDERSON et al., 1973; ÖBERG et al., 1975) are present in large amounts in the empty and incomplete particles. Several additional polypeptides can be identified in the empty and incomplete particles as compared to complete virions. It has been suggested that the empty particles are precursors to virions (SUNDQUIST et al., 1973b)

and virions (PRAGE et al., 1972; WADELL et al., 1973), suggesting that they contain different amounts of DNA. These structures are referred to as *incomplete particles*. Analysis of their polypeptide composition has shown that they also contain precursor polypeptides (Fig. 16). The heavier particles contain less of the precursors than the lighter particles. One of the core proteins, polypeptide V, appears only in particles which at least contain DNA corresponding to half the adenovirus genome. The other core protein, polypeptide VII, is only observed in particles which contain the normal amount of DNA (PRAGE et al., 1974).

2. Defective Assembly

a) Arginine Starvation

Adenovirus maturation requires normal concentrations of arginine in the growth medium (ROUSE and SCHLESINGER, 1967; RUSSELL and BECKER, 1968). Without arginine the virus yield is reduced 3—4 orders of magnitude. This dramatic reduction is somewhat surprising, since the structural proteins, including those rich in arginine, are still synthesized in amounts which would seem to be sufficient to give normal virus production (EVERITT et al., 1971; ROUSE and SCHLESINGER, 1972). When arginine is added after starvation, virus production starts after a lag period of 4—5 hours and an almost normal yield of virus is obtained. However, the capsid proteins made during arginine starvation do not seem to be utilized for assembly of virions after reversion (EVERITT et al., 1971; ROUSE and SCHLESINGER, 1972). The accumulation of virions after reversion is sensitive to inhibition of DNA, RNA and protein synthesis again emphasizing a dependence upon newly synthesized macromolecules. Almost normal yields of virus are obtained when arginine starvation is begun later than 14 hours after infection (EVERITT et al., 1971). This would imply that some critical factor(s), whose synthesis is strongly dependent upon the presence of arginine in the medium, is made earlier than this time. If this factor is absent, the virus components do not assemble. The factor(s) involved has not yet been identified. Starvation reduces the synthesis of viral DNA, whereas all late viral mRNA sequences appear to be synthesized (RASKA et al., 1972). Although large amounts of hexon polypeptide are synthesized in the absence of arginine, there is a 2—4 fold reduction in synthesis of hexon as compared to the other structural proteins (EVERITT et al., 1971). It is not clear if this has relevance for the defective assembly. This effect is similar to the effect of low concentrations of cycloheximide on synthesis of adenovirus polypeptides (PERLMAN et al., 1972).

WINTERS and RUSSELL (1971) reported that viral DNA synthesized in arginine deficient medium can assemble into virions, when mixed with extracts from infected cells maintained in a normal medium. If this finding is confirmed, the system would be useful in elucidating the role of arginine in assembly and to study the route of virus assembly.

b) Elevated Temperature

The synthesis of infectious virus particles is also reduced 2 orders of magnitude if the infected cells are maintained in suspension culture at 42° C. Viral DNA and viral mRNA are synthesized at a faster rate at 42° C, but no virions are formed (WAROCQUIER et al., 1969; OKUBO and RASKAS, 1971). It appears that the viral polypeptides are formed but that the assembly of the polypeptides into capsomeres is defective and this probably causes the decrease in virus yield (OKUBO and RASKAS, 1971).

Using somewhat different experimental conditions PERLMAN et al. (1972) found that translation is altered at 40° C as compared with normal conditions. At 40° C the synthesis of polypeptides is reduced but the degree of reduction varies for the different polypeptides. Normally the ratio between rate of synthesis of hexon and fiber polypeptide is about 4 but may increase to 10 at 40° C.

These results add further support to the idea that the translation of viral mRNA is under complex control.

The assembly may thus involve several steps which may be controlled differently: Rate of synthesis of individual polypeptides, assembly of capsid polypeptides into capsomeres, assembly of capsomeres into empty capsids and subsequently into virions. The transport of newly synthesized polypeptides or capsomeres into the nucleus may be another event which causes defective assembly.

VI. Abortive Infections

Adenoviruses can give rise to productive or abortive infection depending on the serotype and the host cell. Table 5 shows the responses of different cells to infection with different adenovirus serotypes.

Table 5. *Permissiveness for Human Adenovirus Replication in Cells from Different Species*

Species	Cell	Serotype	Permissiveness [a]	Defective event
Human	KB or HeLa Primary fibroblasts or epithelial	All human	+	None
Monkey	AGMK Rhesus	Most human	− or ± [b]	Translation
Hamster	Primary BHK-21 NIL-2	ad 2	+	None
		ad 5	+	None
		ad 12	−	DNA replication
Rat	Primary	ad 2	− or ± [c]	? DNA replication
		ad 12	−	

[a] + denotes production of progeny virions and − indicates that production of virions is deficient.
[b] The permissiveness may vary between human adenovirus serotypes.
[c] The permissiveness may vary among different preparations of rat cells.

A. Adenovirus Infection in Hamster Cells

Hamster cells are permissive for ad 2 and ad 5 whereas ad 12 is unable to replicate in these cells. The ad 12 virus particles can attach to and penetrate baby hamster kidney cells (BHK-21). At least part of the early genes are expressed and T-antigen synthesis can be detected (STROHL et al., 1967; RASKA and STROHL, 1972). However, the ad 12 genome is not replicated (DOERFLER, 1968, 1969; DOERFLER and LUNDHOLM, 1970) and in mixed infections with ad 2 only the latter DNA is able to replicate. There is no transcription of late genes and thus no synthesis of structural proteins.

By following the fate of radioactive parental DNA, DOERFLER (1968, 1969) obtained evidence that a small percentage of the parental ad 12 DNA becomes integrated into the host genome after infection at high multiplicities. The interpretation of these experiments was challenged by ZUR HAUSEN and SOKOL (1969) who, with a different hamster cell line, Nil-2, showed that a significant amount of the parental virus DNA was degraded and reutilized for synthesis of cellular DNA. Subsequently, however, DOERFLER (1970) reported that integration of viral DNA occurs in the absence of both protein and DNA synthesis. Hybridization experiments also showed that viral sequences could be detected at the buoyant density of host DNA when the cells had been prelabeled with BUdR (DOERFLER, 1970). The differences in the results obtained in these investigations could be due to variations between the two cell lines with regard to degradation and reutilization of the parental virus DNA. Most ad 12-infected BHK-21 cells show extensive chromosome pulverization with concomitant degradation of the host cell DNA and these cells finally die (STROHL, 1969a and b). A small population survives the infection, and some of these exhibit altered growth properties characteristic of transformation (STROHL et al., 1970).

In BHK cells arrested in the G1 phase ad 12 induces a rapid increase in cellular DNA synthesis (STROHL, 1969a and b; ZIMMERMAN et al., 1970) with a simultaneous activation of the enzymes involved in DNA replication. The stimulation of DNA synthesis under these conditions is blocked by the addition of cyclic AMP, which also suppresses the synthesis of T-antigen (ZIMMERMAN and RASKA, 1972).

The abortive response of hamster cells to ad 12 infection poses many interesting questions. Do these cells contain a specific inhibitor for ad 12 replication or is the ad 12 genome more defective than ad 2 and requires certain host helper functions which are not present in hamster cells?

Permissive infection of human cells with ad 12 causes a distinct break or gap in the long arms of chromosome No. 17 (ZUR HAUSEN, 1967), a finding which has been utilized to locate the thymidine kinase gene in human chromosomes (MCDOUGALL et al., 1973). This chromosomal aberration is not present in non-permissive or transformed cells, possibly due to the fact that heteroploid lines of hamster cells like BHK and Nil-2 have mainly been used for transformation with this virus (for a review see ZUR HAUSEN, 1973).

B. Adenovirus Infection in Monkey Cells

Monkey cells are semi-permissive for several human adenoviruses (RABSON et al., 1964b). The virus particles enter the cells and the early phase of the infection appears to proceed normally (FELDMAN et al., 1966; FRIEDMAN et al., 1970). In contrast to the abortive response described above, the viral genome is able to replicate in these cells (RABSON et al., 1964b; REICH et al., 1966), but very little progeny virus is produced. Hybridization competition experiments have failed to detect any differences between RNA produced in late lytic infection in KB cells and RNA produced in African green monkey kidney (AGMK) cells; even the virus-induced VA RNA seems to be present (BAUM et al., 1968; FOX and BAUM, 1972; LUCAS and GINSBERG, 1972a). However, little or no synthesis of capsid proteins can be detected (FRIEDMAN et al., 1970; BAUM et al., 1972) and the translation of the late mRNA thus appears to be impaired.

It was discovered some years ago that the block in the replication of adenovirus in monkey cells can be overcome by coinfection with SV 40 virus (O'CONOR et al., 1963; RABSON et al., 1964b). Other viruses, like the simian adenoviruses

(NAEGELE and RAPP, 1967; ALTSTEIN and DODONOVA, 1968), the adeno-SV40 hybrid viruses (ROWE and BAUM, 1965) and the unidentified agent MAC, can also act as helper viruses (BUTEL and RAPP, 1967). In the presence of SV40 helper virus the virus yield is increased 2—3 orders of magnitude. Since the helper SV40 virus is unable to replicate its own DNA under these conditions, it appears likely that an early SV40 function aids the translation of late adenovirus mRNA (FRIEDMAN et al., 1970).

HASHIMOTO et al. (1973) have further investigated adenovirus translation in monkey cells with or without the SV40 helper function. They observed that most of the large adenovirus mRNA (26S) fails to become associated with the polysomes in the absence of helper. This class of adenovirus RNA may either lack a critical sequence for initiation in the absence of helper or the monkey cells lack a critical initiation factor for translation of viral 26S mRNA. FOX and BAUM (1974) also report that the adenovirus mRNA is not readily associated with the polyribosomes although it is polyadenylated normally during infection of AGMK cells. They also proposed that VA RNA was not associated with the polysomes in AGMK cells infected with adenoviruses (BAUM and FOX, 1974).

VII. Adeno-SV40 Hybrid Viruses

The adenovirus-SV40 hybrids are recombinants which contain all or part of the SV40 genome plus a partially deleted adenovirus genome in an adenovirus capsid. Such hybrid viruses were originally isolated when adenovirus was propagated in monkey cells for vaccine production. Stocks of adenovirus serotypes 1—5 and 7 which had been adapted to grow in rhesus monkey kidney cells were found to be contaminated with SV40 virus (HARTLEY et al., 1956). The contaminating SV40 virus was eliminated with SV40-antiserum but the resulting stock had some unusual properties; it gave rise to productive infection in monkey cells, was neutralized with adenovirus but not with SV40 antisera, but still induced SV40 specific T-antigen (RAPP et al., 1964). This strain was designated E46+ or PARA (Particles Aiding Replication of Adenovirus) and was shown to contain particles with both SV40 and adenovirus DNA enclosed in an adenovirus type 7 capsid. The original PARA strain contained two types of particles; wild type, ad7 and particles containing both SV40 and ad7 genetic material (ROWE and BAUM, 1964, 1965). Figure 17 shows schematically a comparison between SV40, ad 7 and PARA particle replication in monkey kidney cells. The SV40 and adenovirus DNA segments present in the hybrid particles are covalently linked since they both band at the density of adenovirus 7 DNA in alkaline CsCl although SV40 DNA has a different buoyant density than ad 7 DNA (BAUM et al., 1966). KELLY and ROSE (1971) have analyzed heteroduplex molecules between E46+ and adenovirus 7 DNA by electron microscopy. The results show that the hybrid contains DNA corresponding to about 75 per cent of the SV40 genome inserted in the adenovirus DNA at 0.05 fractional length from one end of the hybrid DNA. Approximately 10 per cent adenovirus DNA is deleted from the hybrid chromosome and it was suggested that the hybrid DNA arose through two recombination events: one caused the insertion of SV40 DNA and the second event deleted some SV 40 and adenovirus DNA (KELLY and ROSE, 1971). The PARA-strain plaques with two hit kinetics on monkey cells because the hybrid particle is defective and unable to replicate without helper ad 7. On the other hand, wild type ad 7 replicates poorly in monkey cells without the SV 40 helper function which is furnished by the hybrid genome. The progeny of plaques from monkey cells consists of a mixture of ad 7 and hybrid virus. In human cells the hybrid is

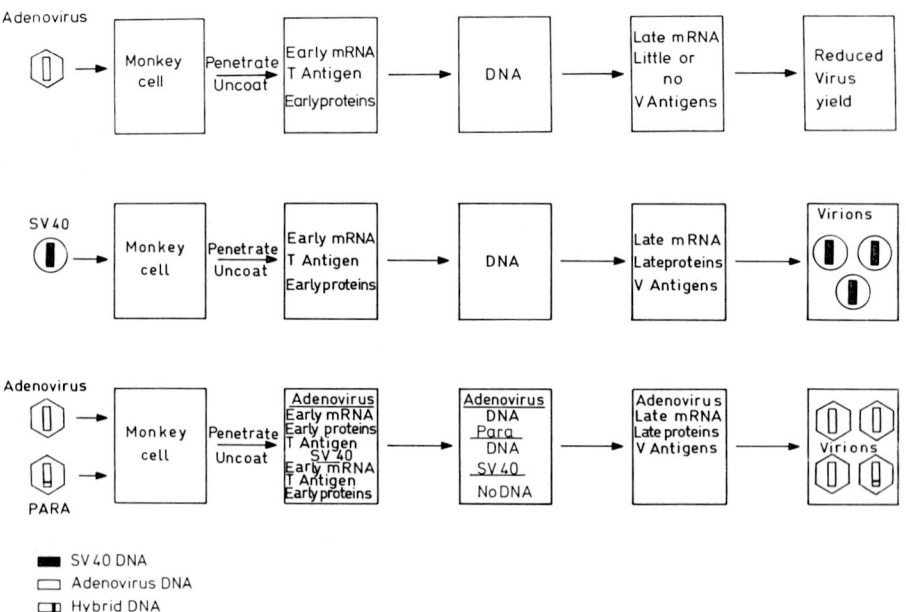

Fig. 17. A diagrammatic representation of events during replication of adeno, SV 40 and the adenovirus-SV 40 hybrid (PARA) viruses in monkey cells.

The molecular events during infection of monkey cells with adenovirus, PARA or SV 40 are shown. Adenovirus reproduction is abortive in monkey cells without simultaneous infection with SV 40 or PARA probably because of a block at the translational level (HASHIMOTO et al., 1973; BAUM and FOX, 1974). An early SV 40 protein appears to overcome this block

unable to replicate without the ad 7 helper and therefore the hybrid is lost when grown at low multiplicities of infection. Synthesis of SV 40 T- and U-antigen is induced by the hybrid, but no SV 40 capsid proteins are synthesized.

The hybrid genome can be transferred to the capsid of a different adenovirus serotype, (transcapsidation) by cocultivation of PARA and other adenoviruses (RAPP et al., 1965; ROWE, 1965). The transcapsidants are insensitive to ad 7 antisera but induce ad 7 T-antigen (ROWE and PUGH, 1966). Transcapsidation between PARA and adenovirus serotypes 1, 2, 5, 6 and 12 has been reported (ROWE, 1965; RAPP et al., 1968).

Adenovirus-SV 40 hybrids have also been isolated from stocks of ad 4 (EASTON and HIATT, 1965; BEARDMORE et al., 1965), ad 5 (LEWIS et al., 1966) and ad 12 (SCHELL et al., 1966). RAPP (1973) recently reviewed this field of research.

Hybrids of special interest have been obtained after propagation of ad 2 in monkey cells. The progeny from the original strain (ad 2^{++}) gives rise to wild-type ad 2, complete SV 40 virions, and a mixed population of adenovirus-SV 40 hybrids (LEWIS et al., 1969). Several stable hybrids have been selected from this

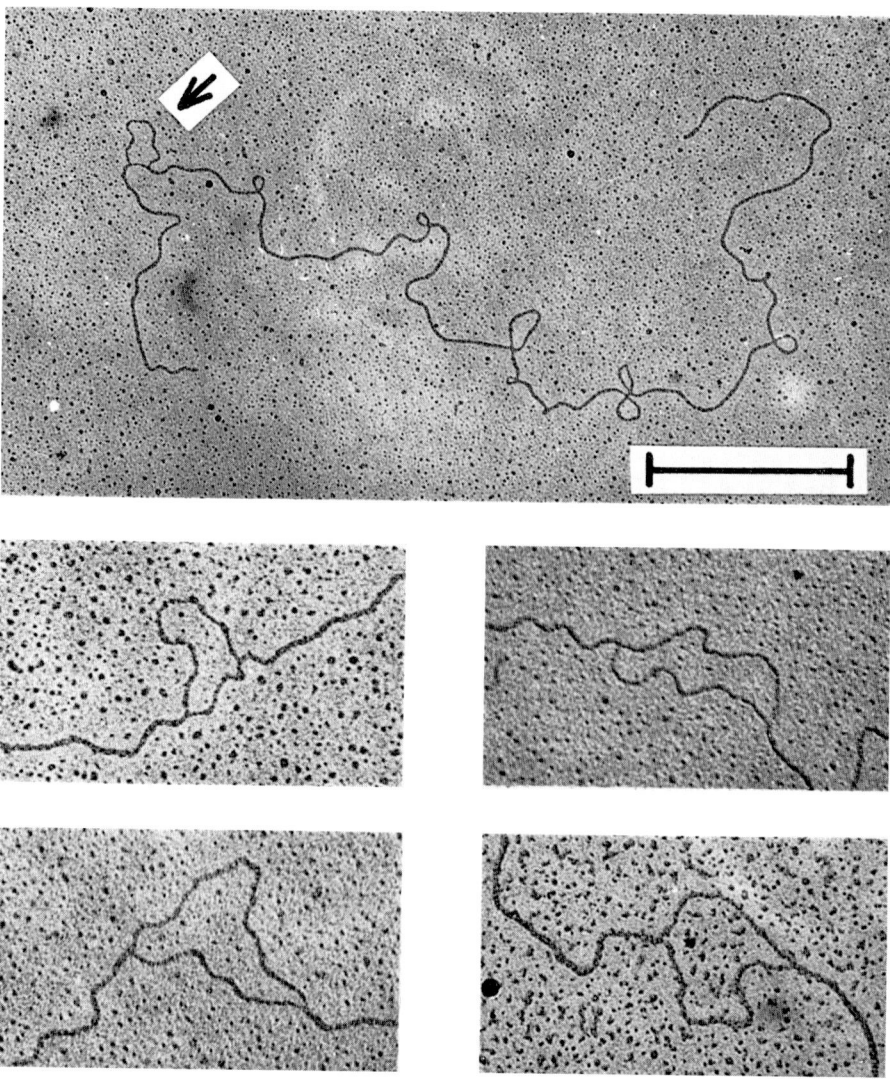

Fig. 18. Heteroduplex molecules between ad 2 DNA and DNA from the nondefective ad 2-SV 40 hybrid viruses.
Top: ad 2/ad 2^+ND_3 heteroduplex. Arrow indicates the position of the unmatched region. The length of the bar corresponds to 1 μm
Bottom (clockwise from top left): Heteroduplex molecules of ad 2/ad 2^+ND_1, ad 2/ad 2^+ND_2, ad 2/ad 2^+ND_4, and ad 2/ad 2^+ND_5. These loops are located in the same region as that of the ad 2/ad 2^+ND_3 heteroduplex
Magnification in the lower frames is twice that of the top frame. This Figure was kindly provided by Dr. T.J. Kelly and is reprinted from KELLY and LEWIS (1973)

strain after propagation in AGMK cells. Two such segregants, known as ad2+ HEY and ad2+ LEY, both plaque with two-hit kinetics on AGMK cells and produce infectious SV40 virus in addition to hybrid particles. They apparently contain the complete SV40 genome covalently linked to a defective adenovirus genome (LEWIS and ROWE, 1970; CRUMPACKER, et al. 1970). For unknown reasons, HEY produces about 10^4 times more infectious virus than LEY (LEWIS and ROWE, 1970).

LEWIS and coworkers (1969) also isolated other hybrid viruses from the ad2++ stock which were non-defective (ND) with regard to replication both in human and monkey cells. The first strain to be characterized was designated ad2+ND$_1$

Table 6. *Properties and Characteristics of Non-Defective Adeno-SV40 Hybrid Viruses*

Strain designation	SV40 DNA segment		% ad 2 genome deleted	Permissive[c] in AGMK	SV40 antigens induced[d]
	Size[a]	% genome[b]			
Ad2+ND$_3$	4.8×10^4	7	5.3	—	None
Ad2+ND$_1$	2.4×10^5	18	5.4	+	U
Ad2+ND$_2$	6.2×10^5	32	6.1	+	U, TSTA
Ad2+ND$_4$	8.4×10^5	43	4.5	+	U, TSTA, T
Ad2+ND$_5$	5.3×10^5	28	7.1	—	None

[a] Amount of SV40 DNA estimated by hybridization of SV40 cRNA to hybrid virus DNA (HENRY et al., 1973). The non-defective adeno SV40 hybrids are listed according to the size of the integrated SV40 fragment.
[b] Estimated by heteroduplex mapping (KELLY and LEWIS, 1973) and refers to the size of the integrated fragment relative to intact SV40 DNA.
[c] Permissiveness indicates that comparable yields of hybrid virus are obtained in human and African green monkey cells (AGMK).
[d] The terminology and the characteristics of the U and T antigens have been described by LEWIS et al. (1973). They are all early products of the SV40 virus infection. TSTA refers to SV40-specific transplantation antigen.

and this virus yields almost equal titers and plaques with one-hit kinetics on both human and monkey cells (LEWIS et al., 1969). Unlike the defective hybrids, ND$_1$ remains stable during replication in human cell lines and cells infected with ad2+ND$_1$ produce the SV40 U-antigen; this antigen is detectable with antisera from hamsters carrying SV40-induced tumors (LEWIS and ROWE, 1971), and resembles the SV40 T-antigen, but is more heat stable. The structure of the ND$_1$ genome has been studied by KELLY and LEWIS (1973) (Fig. 18). Heteroduplex molecules between ad2 and ad2+ND$_1$ DNA were analyzed in the electron microscope and were found to have an insertion of SV40 DNA corresponding to 18 per cent of the SV40 genome. The insertion starts at 0.14 fractional length from one end of the hybrid DNA and at this position a fragment of 1.3×10^6 daltons of adenovirus DNA has been deleted (KELLY and LEWIS, 1973). The SV40 segment in ad2+ND$_1$ contains a control region for SV40 transcription *in vivo* (PATCH et al., 1972) and also a strong promotor for *E. coli* RNA polymerase (ZAIN et al., 1973).

Four additional non-defective ad2-SV40 hybrids designated ad2+ND$_2$ to ad2+ND$_5$ have been isolated (LEWIS et al., 1973) and their properties are summarized in Table 6. HENRY et al. (1973) showed by RNA-DNA hybridization that these hybrids contain different amounts of SV40 DNA covalently linked to ad2

DNA. KELLY and LEWIS (1973) have further analyzed the structure of the non-defective hybrid genomes by heteroduplex mapping techniques and these investigators came to the following conclusions: All non-defective hybrids from strain ad2^{++} contain overlapping segments of SV40 DNA with a common endpoint. Ad2^+ND$_4$ contains the largest insertion corresponding to 40 per cent of the SV40 chromosome and ad2^+ND$_3$ the smallest, about 7 per cent of SV40. In the region where SV40 DNA is inserted, adenovirus DNA has been deleted and the different hybrids have overlapping deletions with a common endpoint at 0.14 fractional length from one of the ends of the hybrid chromosome. Ad2^+ND$_5$ has the largest deletion corresponding to 1.6×10^6 daltons of ad2 DNA. With the exception of ad2^+ND$_5$, it has been shown that there is a positive correlation between the amount of SV40 DNA and the number of detectable SV40 functions after infection with different hybrids. Cells infected with ad2^+ND$_3$ exhibit no detectable SV40 antigen, whereas ad2^+ND$_1$ induces U-antigen, ad2^+ND$_2$ U-antigen and TSTA and ad2^+ND$_4$ induces U-antigen, TSTA and SV40 specific T-antigen. For unknown reasons, ad2^+ND$_5$ fails to induce SV40 specific functions although it contains 28 per cent of the SV40 genome. Possibly it is related to the fact that an unusually large segment of adenovirus DNA was deleted from this hybrid, and the finding that RNA transcribed from the hybrid DNA *in vivo* appears to be controlled by the adenovirus genome (PATCH et al., 1972, 1974). Hybridization competition experiments have shown that only early SV40 RNA is synthesized in cells infected with hybrid viruses (LEVINE et al., 1973). Since the different hybrid viruses induce different SV40 functions, a functional map of the early region of the SV40 genome can be constructed (Table 6). It should be pointed out that no proof is yet available that the SV40 specific antigens induced by different hybrids are separate proteins or even that they are coded for by the hybrid genome.

VIII. Adenovirus Genetics

The first adenovirus mutants were isolated from stocks of ad12 (TAKEMORI et al., 1968). These so called cytocidal (*cyt*) mutants were spontaneously encountered at frequencies of 0.01 per cent but UV irradiation increased the frequency about 5-fold. The mutants can be distinguished because they produce large clear plaques, whereas wild type ad12 produces small fuzzy-edged plaques on human embryonic kidney cells. These *cyt* mutants are less oncogenic for newborn hamsters and have lost the ability to transform hamster cells *in vitro* (TAKEMORI et al., 1968). Recombinants which produce plaques of wild type morphology and which have regained their oncogenicity have been obtained (TAKEMORI, 1972).

Isolation of temperature-sensitive mutants (*ts*) of adenoviruses was first reported in 1971. At present, there is a rapid accumulation of such mutants and a new nomenclature has recently been proposed (GINSBERG et al., 1973). WILLIAMS et al. (1971) and ENSINGER and GINSBERG (1972) have described a large number of temperature-sensitive mutants of ad5. WILLIAMS et al. (1971) isolated mutants after treatment with nitrous acid, nitrosoguanidine, hydroxylamine and bromodeoxyuridine. Mutation frequencies ranged between 0.6—9.6 per cent and the mutants showed little leakiness. The reversion frequencies were in the order of $10^{-5}-10^{-6}$. Fourteen complementation groups have so far been reported by complementation analysis (WILLIAMS and USTACELEBI, 1971; WILKIE et al., 1973; RUSSELL et al., 1972a), but recent results (WILLIAMS, personal communication) seem to expand this figure to at least 17. As mentioned in a previous section, the adenovirus genome could theoretically code for 25—50 average sized polypeptides. Recombination to wild type can be obtained at frequencies ranging from 0.05 to 9.0 per cent (WILLIAMS and USTACELEBI, 1971). ENSINGER and GINSBERG (1972) described 8 ts mutants of ad 5 which were separated into three complementation groups. These mutants also showed low reversion frequencies. Recombination to wild type was observed at frequencies between 0.1—15 per cent. Temperature-sensitive mutants of ad 12 have been isolated by LUNDHOLM and DOERFLER (1971) and SHIROKI et al. (1972). The latter group isolated 88 ts mutants which could be divided into 13 complementation groups. The isolation of ts mutants of adenovirus type 31 (SUZUKI and SHIMOJO, 1971) and avian adenovirus (CELO) has also been reported (ISHIBASHI, 1970, 1971). ISHIBASHI (1971) isolated 49 mutants of the avian CELO virus, which were grouped according to their ability to transport virus antigen from the cytoplasm to the nucleus at the non-permissive temperature.

The ad 12 ts mutants of SHIROKI et al. (1972) and the ad 5 ts mutants of WILLIAMS et al. (1971) have been most extensively studied. SHIROKI et al. (1972) have divided the 13 complementation groups of ad 12 ts mutants into four classes. The first class contains the complementation groups A—D. These mutants are unable to synthesize any of the major capsid proteins, hexon, penton base and fiber at the non-permissive temperature. The second class consists of the mutants in complementation groups E—G which are defective in production of at least two of the major capsid proteins, whereas the third class comprises mutants in the complementation groups H, I and J which are unable to produce either hexon (H), fiber (I) or penton base (J). The mutants in the fourth class which includes the remaining complementation groups K—M produce capsid proteins in normal amounts but are still unable to produce virions at the non-permissive temperature. All these ad 12 mutants are able to synthesize viral DNA and to induce T-antigen, and all are oncogenic for hamsters (SHIROKI et al., 1972).

The ad 5 mutants of WILLIAMS et al. (1971) have been grouped into four classes based on defects in the capsid proteins which can be identified with immunological methods (RUSSELL et al., 1972a). The largest class, class 4, which includes at least 50 per cent of all ad 5 mutants is defective in production of hexon antigen. This class could be subdivided into two subclasses by immunofluorescence; class 4 A mutants are defective in synthesis of the hexon polypeptide whereas class 4 B mutants synthesize this polypeptide but are unable to transport the hexon polypeptide to the nucleus. It is of interest that the serological class 4 consists of at least 6 different complementation groups, which suggests that mutations at several loci of the adenovirus genome affect the synthesis of hexon. Mutants in class 3 comprise 3 complementation groups which all are defective in fiber production. So far only one complementation group of class 2 has been detected (WILKIE et al., 1973) and these mutants are defective in viral DNA synthesis at the non-permissive temperature, but are able to synthesize the early

P-antigen. Finally, class 1 mutants which comprise two complementation groups synthesize DNA and the capsid proteins in normal amounts, but fail to produce intact virions. Two mutants of ad 5 which complement each other fail to induce interferon in chick embryo cells (USTACELEBI and WILLIAMS, 1972). Most of the mutants of ENSINGER and GINSBERG (1972) were defective in hexon synthesis. Mutants in one of their complementation groups are unable to synthesize DNA (ts 125), while members of the third complementation groups synthesize DNA and capsid proteins but are defective in some unknown function (ENSINGER and GINSBERG, 1972).

Available results only allow a crude classification of adenovirus conditional lethal mutants. However, the rapid accumulation of temperature-sensitive mutants of different adenovirus serotypes should soon give a detailed genetic map of the adenovirus chromosome. GRODZICKER et al. (1974) recently described recombinants between ts mutants of ad 2^+ND_1 and ts mutants of ad 5 which were used to relate the genetic map of ad 5 obtained by two factor-crosses with the physical map obtained by restriction endonucleases. Since ad 5 and ad 2 DNA have different cleavage sites for the EcoRI endonuclease (Table 1), the recombinants should exhibit an intermediate cleavage pattern. The results suggest that the ad 5 ts-mutants are distributed over the entire genome and that the fiber mutants are located in the EcoRI-C fragment of ad 2 (GRODZICKER et al., 1974). This procedure may be used to create a detailed genetic map.

IX. Cell Transformation

The human adenoviruses have been divided into highly, weakly and non-oncogenic serotypes corresponding to the subgroups A, B and C, respectively (HUEBNER, 1967). This subdivision was originally based on the frequency with which hamsters develop tumors after inoculation with virus. It was later established that several human adenoviruses and adenoviruses from other species could transform cells *in vitro* in some cases irrespective of oncogenic capacity *in vivo*. Human adenoviruses like most papovaviruses transform most frequently cells which are non-permissive for virus replication. Recently, however, WILLIAMS (1973) observed that a temperature-sensitive mutant of ad 5 can transform permissive cells at a temperature which does not permit virus replication. This may suggest that the events leading to transformation also occur in permissive cells, but is not detected because the infected cells do not survive.

Cell transformation with adenoviruses requires high multiplicities of infection and the frequency of transformation is low. The transformed cells, which can be selected because of their unlimited growth potential, have a characteristic morphology and BHK-21 cells transformed by ad 12 and polyoma virus can be distinguished (STROHL et al., 1967). The cells grow to higher saturation densities than untransformed cells, grow in disoriented arrays and have an altered plasma membrane (MARTINEZ-PALOMO and BRAILOVSKY, 1968). Infectious virus cannot be detected in transformed cells, and infectious virus has never been induced from adenovirus-transformed cells by forming heterokaryons with permissive cells or by treating the cells with physical or chemical agents (BURNS and BLACK, 1969a;

WEBER, 1974), which induce virus production in papovavirus-transformed cells (GERBER, 1966; BURNS and BLACK, 1969b; FOGEL and SACHS, 1969, 1970). In spite of the failure to rescue virus production, several lines of evidence indicate that at least parts of the adenovirus genome persist in transformed cells.

BURNETT and HARRINGTON (1968b) reported that naked DNA of simian adenovirus type 7 (SA7) can induce tumors and MAYNE et al. (1971) demonstrated that fragmented SA7 DNA also was tumorigenic. Subsequently, GRAHAM and VAN DER EB (1973b) have demonstrated that cell transformation can be obtained with ad5 DNA in the presence of calcium phosphate. Fragmented DNA can also cause cell transformation and a DNA segment of about 1×10^6 daltons appears to be sufficient for transformation. This method opens new possibilities to obtain information about adenovirus transformation, since it should be possible by this method to identify the DNA sequences which are necessary for transformation. GRAHAM et al. (1974) have recently been able to transform cells with a DNA fragment which originates from the left end of the ad5 chromosome (Fig. 12).

A. Cell Transformation by Different Adenoviruses

The subgroup A adenoviruses were first shown to transform newborn hamster kidney cells (McBRIDE and WIENER, 1964). FREEMAN et al. (1967a) subsequently reported transformation of rat embryo fibroblasts with ad12. Both hamster and rat cells are non-permissive for ad12 replication (LEVINTHAL and PETERSON, 1965). Transformation is more reproducible if the cells are kept in growth medium with low calcium and under such conditions in vitro transformation of rat embryo is also observed with serotypes 1 and 2 from the non-oncogenic subgroup C (FREEMAN et al., 1967b; McALLISTER et al., 1969a). Rat cells transformed by these viruses are, however, not tumorogenic unless the animals are immunosuppressed (GALLIMORE, 1972). In addition, ad2 and the nondefective adenovirus 2-SV40 hybrid can transform permissive hamster cells and the transformed cell lines have been shown to produce tumors when injected into suckling hamsters (LEWIS et al., 1974). Serotypes of subgroup D, which also includes non-oncogenic adenoviruses, can also transform rat and hamster cells in vitro (McALLISTER et al., 1969b). Simian adenoviruses, SA7 and SA11 can transform hamster cells in the same way as human ad12 (RIGGS and LENETTE, 1967; CASTO, 1968a; WHITCUTT and GEAR, 1968; ALTSTEIN et al., 1967; CASTO, 1969).

B. Viral DNA in Transformed Cells

Evidence has been obtained that ad12 DNA becomes integrated into host DNA during infection of non-permissive hamster cells (DOERFLER, 1970) and some of these cells eventually become transformed. It is therefore likely, although not proven, that viral DNA is integrated in all adenovirus transformed cells. Viral DNA has been detected in transformed cells by the method of WESTPHAL and DULBECCO (1968). RNA transcribed in vitro by E.coli RNA polymerase with adenovirus DNA as a template (cRNA) has been shown to specifically hybridize to DNA extracted from cells transformed with the homologous adenovirus (GREEN et al., 1970). Quantitative estimates using filter hybridization techniques showed

14—37 copies of viral DNA in ad2 transformed rat cells, and 22 and 97 copies in hamster cells transformed by ad12 and ad7, respectively. With the same technique, DOERFLER (1970) found 5—60 equivalents of adenovirus DNA in ad12 hamster cells at 30—40 hours after non-permissive infection. The filter hybridization assay of WESTPHAL and DULBECCO (1968) has sometimes given inaccurate results because this method relies on reconstruction experiments with known amounts of viral DNA on the filters. Errors arise because hybrids are specifically lost from the filters in such reconstruction experiments (HAAS et al., 1972). A less ambiguous method to quantitate viral DNA in transformed cells has been designed by GELB et al. (1971). The rate of renaturation of ^{32}P-labeled viral DNA in the presence and absence of DNA from transformed cells is followed by hydroxylapatite chromatography. By this method, PETTERSSON and SAMBROOK (1973) found one copy of viral DNA per diploid quantity of cell DNA in one line of ad2 transformed rat cells. More detailed analysis of the ad2 DNA present in this line of transformed cells with specific fragments of ad2 DNA (PETTERSSON et al., 1973) indicated that close to two copies of about 45 per cent of the viral genome was present (SHARP et al., 1974b). Sequences corresponding to Endo R. Eco RI fragments C, D, E and part of A were detected. When additional cell lines of ad2 transformed rat cells were analyzed, it has been found that all transformed cell lines contain viral genomes with deletions. GALLIMORE et al. (1974) have examined the viral genome in nine ad2 transformed cell lines, seven of which were independently transformed. The results showed that four of the independently transformed lines and the two sublines contained an identical segment of viral DNA. This segment stretches from the left hand end to a point about 14 per cent along the viral genome (Fig. 12). The three remaining lines contained this segment and additional sequences from other parts of the viral genome. In agreement with these findings, GRAHAM et al. (1974) have found that a fragment of naked ad5 DNA from the left hand of the genome is sufficient to cause transformation *in vitro*. Thus, all adenovirus transformed cells seem to contain incomplete viral genomes which explains the difficulty in rescuing infectious virus from transformed cells (WEBER, 1974).

In situ hybridization has been used to determine the location of adenovirus DNA in transformed cells. No preferential association of the ad2 DNA with specific chromosomes has so far been observed (MCDOUGALL et al., 1972; LONI and GREEN, 1973).

C. Synthesis of Viral RNA

Since the adenovirus genome in transformed cells constitutes a minute fraction of the total cell DNA, the viral transcripts would be expected to amount to a very small fraction of the total RNA in transformed cells. In contrast, FUJINAGA and GREEN (1966) found that as much as 2 per cent of the mRNA in the cytoplasm of ad12 transformed hamster cells was of viral origin. This implies that the viral DNA is preferentially transcribed in transformed cells. By competition hybridization experiments it has been established that only early RNA sequences are transcribed in ad2 transformed rat cells and that about 50 per cent of the early sequences can be detected in these cells (GREEN et al., 1970). In ad7 transformed hamster cells, on the other hand, all early RNA sequences seem to be expressed (GREEN et al., 1970).

Viral RNA in cells transformed by adenoviruses from different subgroups have similar base compositions (47—51 per cent GC), although the DNAs from the transforming viruses differ considerably in their base composition (FUJINAGA and GREEN, 1968). In contrast, viral RNA in cells transformed with viruses from subgroup A, B and C, hybridizes only to DNA from members of the same subgroup (FUJINAGA and GREEN, 1968a and b, 1968; FUJINAGA et al., 1969). Thus, RNA transcribed in transformed cells appear to originate from portions of the adenovirus DNA which have subgroup specific base sequences. If a common adenovirus "transforming gene" exists, it has been subject to considerable genetic drift. This does not necessarily mean that the adenovirus gene products present in different transformed cells are different to the same extent. It was, for instance, recently shown that genes coding for histones from different species show considerable sequence heterology although the amino acid sequence of the proteins is remarkably well conserved (BIRNSTIEL et al., 1973).

The viral genome in transformed cells appears to be transcribed into RNA which is similar in size to HnRNA in uninfected cells (GREEN et al., 1970; TSEUI et al., 1972; WALL et al., 1973). Nuclear RNA from transformed cells which is selected by hybridization to adenovirus DNA also hybridizes with cellular DNA, which suggests that the integrated genome is transcribed into RNA molecules which contain covalently linked viral and cellular sequences (TSEUI et al., 1972; WALL et al., 1973). By selecting large viral HnRNA molecules which contained poly (A) from ad2 transformed rat cells, BACHENHEIMER (1974) have provided evidence that the viral specific sequences are located at the 3'-terminus of the HnRNA. Since the mRNA in the cytoplasm is smaller than nuclear HnRNA containing viral sequences, it appears that the viral RNA in transformed cells is subject to the same processing mechanism as HnRNA in uninfected cells (WALL et al., 1973). Cytoplasmic RNA from transformed cells contains two main size classes of RNA sedimenting at 16 and 20S. Thus, virus-specific RNA in transformed cells is similar in size to the viral mRNA in early productive infection (WALL et al., 1973; LINDBERG et al., 1972).

In conclusion, it appears that only a minor part of the viral genome is expressed in adenovirus transformed cells and that the viral RNA transcribed in cells transformed by adenoviruses from subgroups A, B and C have a similar base composition but do not show detectable homology by hybridization. Like cells transformed with papovavirus, adenovirus-transformed cells provide a useful system for studying the mechanism of mRNA synthesis in eukaryotic cells.

D. Properties of Adenovirus Transformed Cells

Originally, ad 12 transformed cells were noted as characteristic foci of epithelial-like cells which tend to pile up in clusters (McBRIDE and WIENER, 1964). Similar observations were reported by others for adenovirus-transformed rat and hamster cells (POPE and ROWE, 1964; STROHL et al., 1966; FREEMAN et al., 1967a and b; KUSANO and YAMANE, 1967; RAFAJKO, 1967). The transformed cells have an increased content of DNA, which can be detected with Feulgen staining (KUSANO and YAMANE, 1967). Adenovirus-transformed cells also show an increased amount of membrane bound mucopolysaccharides as has also been found after transformation by papovaviruses (MARTINÉZ-PALOMO and BRAILOVSKY, 1968). The frequency of transformation is extremely low and transformation occurs during the course of many months. Factors like the growth state of the cells at infection, the multiplicity of infection as well as the composition of the medium influence the efficiency of transformation (SCHELL and SCHMIDT, 1968; SCHELL et al., 1968a and b). Under favourable conditions, colonies of tightly packed cells can be identified 2 to 3 weeks after inoculation. Cell transformation requires usually about 10^4-10^6 times more particles than lytic infection in permissive systems (for a review see CASTO, 1973). Normal levels of calcium seem to be essential for transformation, but the establishment of cell lines usually requires a reduction of the calcium concentration in the growth medium to 0.1 mM (FREEMAN et al., 1967a; SCHELL et al., 1968a; VAN DER NOORDA, 1968a and b). The frequency of transformation by ad 12 has been studied by STROHL et al. (1967). Cloned BHK-21 cells were infected at a high multiplicity of infection and assayed for colony growth in soft agar. After three weeks 0.1–1 per cent of the cells had formed visible colonies. On the other hand, McALLISTER and McPHERSON (1968) reported a transformation frequency of 0.002 per cent for another established hamster cell line, Nil-2, after infection with ad 12 and the same investigators failed to transform BHK-21 cells.

Although adenovirus-transformed cells have a characteristic morphology (KUSANO and YAMANE, 1967) and specific growth properties, it has not been unequivocally proven that the adenovirus genome is responsible for the transformed phenotype. Transformation occurs at low frequencies and spontaneous transformation is sometimes seen in cultures of rodent cells.

Adenovirus-transformed cells contain a specific antigen, the so called T-antigen, which can be detected by complement fixation or immunofluorescence. It is detectable in cells transformed by all subgroups of human adenoviruses (HUEBNER et al., 1963; POPE and ROWE, 1964; HUEBNER, 1967) and it is also synthesized early in productive infection (HOGGAN et al., 1965). There is no cross-reaction between T-antigen from adenovirus and that from papovavirus-transformed cells and the T-antigens from human adenovirus subgroups A, B and C differ immunologically. The "non-oncogenic" adenoviruses have been divided into two groups, C and D, because members of the two groups induce immunologically different T-antigens (McALLISTER et al., 1969b). T-antigens from ad 12-transformed cells have been purified (TAVITIAN et al., 1967; GILEAD and GINSBERG, 1968a and b; TOCKSTEIN et al., 1968) but there is no general agreement on their properties. In addition to the T-antigen, a specific transplantation antigen (TSTA), which is responsible for transplantation rejection, appears on the surface of adenovirus transformed cells (SJÖGREN et al., 1967). At present, there is no direct evidence that the viral genome codes for either TSTA or T-antigen. Surface changes detected as increased lectin agglutinability have been found in productive infection (SALZBERG and RASKAS, 1972), but it is not yet established that the TSTA is present during productive infection.

X. Tumor Induction by Adenoviruses

A. Induction of Tumors in Rodents

As mentioned in a previous section, the human adenoviruses have been divided in groups on the basis of their oncogenicity for newborn hamsters (HUEBNER, 1967). Subgroup A comprises serotypes ad 12, ad 18 and ad 31 which are "highly oncogenic" and subgroup B comprises among other types ad 3 and ad 7, which are "weakly oncogenic" (Table 7). The terms highly, weakly and non-

Table 7. *Classification of Human Adenoviruses*

Subgroup	Serotypes	Per cent GC in DNA[a]	Per cent DNA-DNA homology between types[b]	Hemagglutination subgroup[c]	Length of fiber[d] nm	Free dodecons[d]	Oncogenicity[e]
A	12, 18, 31	48—49	80—85	IV	28—31	—	High
B	3, 7, 11, 14, 16, 21	50—52	70—95	I	9—11	+	Weak
C	1, 2, 5, 6 4	57—59	85—95	III	23—31 17—18	— +	None but transform
D	8, 9, 10, 13, 15, 17, 19, 20, 22 22—28	57—61	N. K.	II	12—13	+	None but transform

N.K. = not known.
[a] Data from PIÑA and GREEN (1965) and GREEN (1970).
[b] Data from LACY and GREEN (1964, 1965 and 1967) and also confirmed by heteroduplex mapping (GARON et al., 1973).
[c] Data from ROSEN (1960).
[d] Data from NORRBY (1969b) and LIEM et al. (1971).
[e] Highly oncogenic types cause tumors within 2 months, weakly oncogenic types cause tumors in some animals in 4—18 months.

oncogenic refer to the frequency of tumor induction in hamsters after inoculation with virus. However, the tumor cells once established show no fundamental difference. The non-oncogenic adenoviruses, many of which can transform cells *in vitro*, have been divided into subgroups C and D (HUEBNER, 1967; McALLISTER et al., 1969b) based on differences in the antigenicity of T-antigen. In addition to the human adenoviruses from subgroups A and B, it has been found that several adenoviruses from animals are oncogenic for newborn hamsters. These viruses include the chick embryo lethal orphan (CELO) virus, (SARMA et al., 1965), several simian adenoviruses (HULL et al., 1965) as well as the adenoviruses associated with infectious canine hepatitis (ICH) (SARMA et al., 1967). Bovine adenovirus type 3 and possibly additional bovine serotypes can also induce tumors in hamsters (DARBYSHIRE, 1966, 1973). All adenovirus tumors have a characteristic cytology which often is referred to as an adenovirus-specific microscopic pattern (SLIFKIN et al., 1968). In hamsters these tumors have been characterized as undifferentiated sarcomas (TRENTIN et al., 1968), lymphosarcomas (LARson et al., 1965), undifferentiated fibrosarcomas (McAllister et al., 1966) as well as primitive undifferentiated mesenchymal neoplasms (Huebner et al., 1962). They are all composed of neoplastic cells which are large and uniform, and the tumors sometimes also contain multinucleated giant cells. Since the adenovirus-induced tumors have an almost identical microscopic morphology irrespective of the location of the neoplasm and the route of inoculation (Slifkin et al., 1968; Yabe et al., 1963), it has been suggested that the morphology of adenovirus neoplastic cells is determined by the viral genome (Strohl et al., 1967). In contrast, the papovaviruses seem to induce neoplasms in hamsters which are more closely related to the character of the target cell (Diamondopoulos, 1968a and b). It is conceivable that the papovaviruses with their small genome are less capable to influence the morphology of the tumor cell whereas the adenovirus genome is sufficiently large to code for functions which influence tumor cell morphology.

Adenovirus-induced tumors appear to be free of infectious virus and attempts to rescue adenovirus infectivity from transformed or tumor cells have been unsuccessful (Burns and Black, 1969a; Weber, 1974, and for a review see Casto, 1973). The expression of adenovirus genes in tumor cells has not yet been subject to a thorough investigation. The tumor cells produce T-antigens which are specific for subgroups A, B, C and D, respectively (Huebner et al., 1963; Pope and Rowe, 1964; Huebner, 1967; McAllister et al., 1969b), and in addition specific transplantation antigens can be demonstrated on the surface of the tumor cells (Trentin and Bryan, 1964; Sjögren et al., 1967; Berman, 1967). Usually, adenovirus structural proteins are not produced in tumor cells, although it has been reported that animals bearing large virus-induced or transplanted tumors produce antibodies, which react with the fiber and the unidentified D-antigen, and which neutralize virus infectivity (Huebner et al., 1964; Berman and Rowe, 1965). Concentrated tumor extracts have occasionally been reported to yield incomplete virus particles which band at a low density in CsCl gradients (Smith and Melnick, 1964). In contrast, studies of single cells by immunofluorescence and electron microscopy have repeatedly failed to detect virions or intranuclear crystals of structural proteins in the tumor cells. It has therefore been concluded that the structural proteins which are occasionally detected in tumor extracts must emanate from a minor population of cells in the tumor.

Hormonal factors may play a role in adenovirus oncogenesis since more female hamsters develop tumors than males (Yohn et al., 1965). For hamsters it was found that about 20 per cent more females develop tumors when low doses of ad 12 were inoculated. With high doses, 95—100 per cent of all hamsters developed tumors within 3—8 weeks irrespective of sex (Green and Piña, 1964; Yohn et al., 1967). A similar preference for female hamsters has been reported for tumors induced by ad 7 and ad 18 among the human adenoviruses (Yohn, 1973), SV 20 and SA 7 among the simian adenoviruses (Yohn, 1973; Hatch et al., 1970) and also for the avian CELO virus (Jones et al., 1970). The tumors can be transplanted with a higher frequency in females, regrow faster after excision, have a shorter latent period, and show less regression (Yohn et al., 1968). This enhanced oncogenesis in female hamsters could not be correlated with a higher frequency of in vitro transformation (Casto, 1968b). It has been ruled out that the difference was due to variation in immunological competence between female and male animals, and evidence has accumulated to suggest that the effect is due to hormonal factors. Accordingly, estrogens seem to enhance adenovirus oncogenesis and androgens may be inhibitory (for a review see YOHN, 1973). This enhanced oncogenesis for females could be used to study hormonal effects on tumor induction.

At least 12 of the 31 human adenovirus types (McALLISTER et al., 1972) can transform rodent cells in vitro and the transformed cells are usually tumorigenic. Hamster cells transformed in vitro by subgroup A and B viruses form tumors when injected into the host (McALLISTER and McPHERSON, 1968; McALLISTER et al., 1969a and b). Recently, it was also reported that the "non-oncogenic" ad 2 and the adenovirus-SV40 hybrids can transform hamster kidney cells, which after transplantation give rise to tumors in hamsters (LEWIS et al., 1974). In contrast, rat cells transformed with ad 2 can only establish tumors in immunosuppressed animals (GALLIMORE, 1972). This could be due to the presence of a strong transplantation antigen on the cell surface of transformed rat cells.

B. Do Adenoviruses Play a Role in Human Cancer?

Human adenovirus type 12 can produce foci of human cells in vitro, which appear as transformed cells. However, no cell lines have been established possibly because of the lytic effect of the virus in this permissive system (TODARO and ARONSON, 1968; SHEVLIAGHYN and KARAZAS, 1970). It appears to be a rule for both papova- and adenoviruses that tumor induction occurs in host cells which are non-permissive for virus replication. Consequently, it may seem unlikely that human adenoviruses play a role in the etiology of human cancer. However, recent developments have somewhat changed the outlook on this problem. It has been found that permissive hamster cells can be transformed by a strain of ad 5 with a temperature-sensitive mutation, which prevents replication of the virus and subsequent cell lysis (WILLIAMS, 1973). LEWIS et al. (1974) have also shown that non-oncogenic ad 2 can transform permissive hamster kidney cells. It has furthermore recently been emphasized that rat cells are semipermissive for replication of ad 2 although transformation occurs in such cells (McDOUGALL et al., 1974).

Cells transformed in vitro by adenoviruses contain several footprints of the viral genome. Adenovirus-specific T-antigen and viral RNA can be detected. Serological and biochemical tools have been used to search for such footprints of adenoviruses in human cancer cells. Although the presence of virus particles in neoplastic tissue has been described (SOHIER et al., 1963; BRONITKI et al., 1964; McALLISTER et al., 1964), biochemical and serological tests have failed to demonstrate any signs of adenovirus activity in human tumor cells. A collaborative investigation sponsored by the National Cancer Institute was arranged to search for complement-fixing antibodies against T-antigen from human subgroups A—D and non-human adenoviruses in cancer patients (GILDEN et al., 1970). No positive results were obtained. With a more sensitive immunofluorescence technique several human sera were found to contain antibodies to adenovirus T-antigen (LEWIS et al., 1967), but a comparison between cancer patients and control subjects revealed no difference (ROWE and LEWIS, 1968; GILDEN et al., 1970). RNA samples from 200 human cancer specimens have also been analyzed by hybridization competition for the presence of adenovirus specific RNA. With DNA from representative serotypes from human subgroups A, B and C as a probe, no viral sequences could be detected (McAllister et al., 1972). Because of the limited sensitivity of the assay, the results only mean that less than 1000 molecules of RNA are present per tumor cell. Since human cells are permissive for subgroups A, B, C and D viruses, it is conceivable that if adenoviruses were a causative agent of human tumors only a fraction of the viral genome would be integrated in the DNA of the tumor cells. Therefore a more sensitive probe may be required to exclude that adenovirus DNA is present in human cancer cells. In addition, the fact that virus-specific RNA could not be detected in ad19 transformed cells (McAllister et al., 1972) may mean that transformation with adenoviruses does not in all cases result in a continuous expression of viral genes.

In conclusion, the results obtained by serological and biochemical tests give no indication that adenoviruses play a role in the etiology of human cancer. It cannot, however, be excluded that more sensitive methods will reveal adenovirus specific products in human tumor cells.

XI. Biochemistry and Immunology of Adenovirus Structural Proteins

Soon after the discovery of the adenoviruses, it was recognized that virus multiplication is accompanied by production of virus-specific antigens in the infected cells (Huebner et al., 1954; Hilleman et al., 1955). The antigens were separated into three classes by immunoelectrophoresis (Pereira et al., 1959) and by DEAE chromatography (Klemperer and Pereira, 1959; Philipson, 1960; Wilcox and Ginsberg, 1961; Haruna et al., 1961). The three classes of antigens were related to the capsid proteins both by antigenicity and structure (Wilcox and Ginsberg, 1963a; Wilcox et al., 1963; Valentine and Pereira, 1965; Norrby, 1966) and were subsequently identified as the hexons, pentons and fibers as defined in the section on the architecture of the virion (section II). In the late phase of adenovirus infection the synthesis of host cell proteins is turned off (Ginsberg et al., 1967) and the infected cells are primarily engaged in making large amounts of viral structural proteins. The polypeptides of the structural proteins are made in large excess. Most of the structural polypeptides assemble into multimeric structural proteins (Horwitz et al., 1969) but only 1 to 5 per cent of the fibers and the penton bases and 20—30 per cent of the hexons are assembled into mature virions (White et al., 1969; Everitt et al., 1971). During recent years interest has focused on the structural proteins of adenovirus since unlike the structural proteins of most other animal viruses they are soluble under non-denaturing conditions. They are furthermore available in sufficient amounts for immunological and chemical characterization. The antigens from the excess pool have been almost the exclusive source for purification of adenovirus proteins. Several methods have been described for growth and purification of the adenovirus antigens, most of which are modifications of the original method of Green and Piña (1963). Separation of the virion from the antigens is usu-

Table 8. *Characteristics of Virion Proteins of Adenoviruses*

	Hexon	Penton base	Fiber	Core protein I	Core protein II	Hexon associated I	Hexon associated II	Hexon associated III
Number per virion	240	12	12	~1000	~200	~450		
Molecular weight	310,000–360,000 [a]	400,000–515,000 [b]	200,000 [c] (ad 2)	17,000 [d]	45,000 [e]	50,000 [f]	15,000 [f]	
Polypeptide size	90,000–120,000 [g, h] 3/hexon	70,000 [g] Presumably 5/penton base	60,000–65,000 [g] Presumably 3/fiber	18,500 [g]	48,500 [g]	24,000 [g]	13,000 [g]	12,000 [g]
Polypeptide designation (see Fig. 2)	II	III	IV	VII	V	VI	VIII	IX
Morphology	8 × 12 nm [i] cylindrical	8 nm sphere [b]	9–31 nm × 2 nm rod	NK	NK	NK	NK	NK
Associated biological activity	NF	Cell detaching, Endonuclease (?)	Inhibition of macromolecular synthesis (?), Hemagglutination, Attachment to cell receptors	NF	NF	NF	NF	NF

Four additional polypeptides (IIIa, X, XI, XII) have been identified on SDS-polyacrylamide gels (EVERITT et al., 1973; ANDERSON et al., 1973) (see Fig. 2).

[a] FRANKLIN et al., 1971a.
[b] PETTERSSON and HÖGLUND, 1969; WADELL, 1970.
[c] SUNDQUIST et al., 1973a.
[d] PRAGE and PETTERSSON, 1971.
[e] LAYER, 1970.
[f] EVERITT and PHILIPSON, 1974.
[g] Estimated from SDS-polyacrylamide gel electrophoresis (EVERITT et al., 1973; ANDERSON et al., 1973).
[h] Crystallographic and chemical methods give values in this range. (CORNICK et al., 1973; JÖRNVALL et al., 1974b; HORWITZ et al., 1970.)
[i] PETTERSSON et al., 1967; TEJG-JENSEN et al., 1972.

NK = not known, NF = none found

ally achieved by ultracentrifugation (GREEN and PIÑA, 1963) or by exclusion chromatography on 4 per cent agarose (PETTERSSON et al., 1967). Group separation of the antigens into the three classes can be carried out by DEAE-chromatography or by exclusion chromatography on Sephadex G 200 (NORRBY and SKAARET 1967). Table 8 summarizes several characteristics of the isolated virion proteins from ad 2.

A. The Hexon

Several methods for purification of adenovirus hexons from ad 2 and ad 5 have been reported. KÖHLER (1965) used methanol precipitation at pH 4.0 and DEAE chromatography. Other investigators (VALENTINE and PEREIRA, 1965;

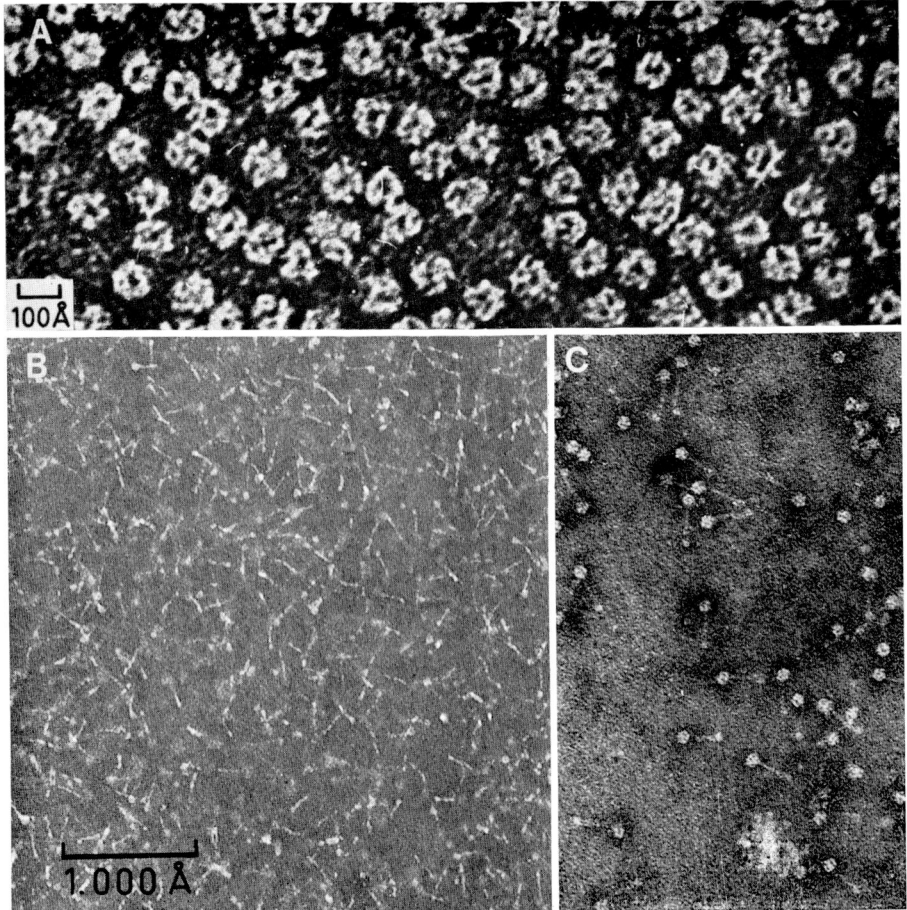

Fig. 19. Electron microscopy of isolated capsid proteins from ad 2.
A. Purified type 2 hexons contrasted with uranyl acetate (PHILIPSON and PETTERSSON, 1973)
B. Purified type 2 fibers contrasted with phosphotungstic acid at pH 7.6 (PETTERSSON et al., 1968)
C. Purified type 2 pentons contrasted with uranyl acetate (PETTERSSON and HÖGLUND, 1969)

HOLLINSHEAD et al., 1967; RUSSELL et al., 1967a; KJELLÉN and PEREIRA, 1968; MAIZEL et al., 1968b) have used repeated cycles of DEAE chromatography as the sole method for hexon purification. Additional steps beyond ion-exchange chromatography seem to be necessary to achieve homogeneity. LEVINE and GINSBERG (1967) combined DEAE and brushite chromatography and PETTERSSON et al. (1967) introduced preparative polyacrylamide gel electrophoresis as an additional purification step to obtain highly purified hexon for biochemical studies. BOULANGER et al. (1969) used liquid film electrophoresis and WADELL (1970) and SHORTRIDGE and BIDDLE (1970) used isoelectric focusing for further purification. Crystallization (PEREIRA et al., 1968) has also been used as an additional purification step for large scale preparation of purified hexon (COUCH et al., 1973). BOULANGER et al. (1973) have recently described a two-step procedure with DEAE and hydroxylapatite chromatography to obtain purified products of all the three classes of antigens.

1. Morphology

Several investigators have studied the morphology of the hexon by electron microscopy. HORNE et al. (1959) described the hexons as solid spheres as did VALENTINE and PEREIRA (1965) in their original paper on the structure of adenoviruses. WILCOX and GINSBERG (1963a) described them as hollow polygonal rods. PETTERSSON et al. (1967) found a cylindrical structure with a diameter of 8—10 nm and with central holes 2.5 nm in diameter (Fig. 19). The dimensions of native type 2 hexons were recently determined by low angle X-ray diffraction studies (TEJG-JENSEN et al., 1972) and the hexons were shown to be cylinder-like objects with a height of 12.5 nm and a diameter of 8—8.5 nm. In preparations of disrupted ad5 virions SHORTRIDGE and BIDDLE (1970) found two separate populations of hexons; one with and one without central holes. They suggested that the holes might be covered with additional polypeptides in some hexons.

2. Physical-Chemical Properties

The sedimentation constant of hexons from ad2 has been determined to be 12.9 (FRANKLIN et al., 1971a). The molecular weight of the hexon has been a matter of controversy and the figures reported in the literature vary between 180,000 and 400,000. Early molecular weight determinations by electron microscopy (VALENTINE and PEREIRA, 1965) and gel filtration (WASMUTH and TYTELL, 1966) suggested a molecular weight between 180,000—230,000 whereas by sedimentation diffusion studies values between 300,000 (KÖHLER, 1965) and 400,000 (PETTERSSON et al., 1967) were obtained. FRANKLIN et al. (1971a) determined the molecular weight for the ad2 hexon by several methods including sedimentation equilibrium and X-ray crystallography and arrived at values in the range of 310,000—360,000 daltons. The amino acid composition of ad2, ad3 and ad5 hexons has been determined (BISERTE et al., 1966; PETTERSSON et al., 1967; BOULANGER et al., 1969; LAVER, 1970 and PETTERSSON, 1971). No major differences can be discerned between hexons from different serotypes. They are all rich in dicarboxylic amino acids (Table 9) and no carbohydrate has been detected (PETTERSSON et al., 1967; BOULANGER et al., 1969). Originally, no half-cystine

Table 9. *Amino Acid Composition of Adenovirions and their Structural Proteins*

Amino acid*	Type 2[a] virions	Type 2[b] hexons	Type 3[b] hexons	Type 5[b] hexons	Type 2[c] fibers	Type 2[d] pentons	Type 2[e] Major core protein	Type 3[e] Major core protein	Type 2[f] protein	Type 2[f] protein	Type 2[f] protein
Polypeptide	—	II	II	II	IV	III + IV	VII	VII	VI	VIII	IX
Lys	4.4	4.2	4.0	4.6	6.0	5.1	3.8	3.7	1.6	1.8	1.8
His	1.6	1.4	1.7	1.7	0.8	1.6	2.2	1.8	2.5	2.4	0.6
Arg	7.9	4.6	4.9	5.0	1.5	3.9	20.6	23.2	6.3	5.5	6.3
Asp	11.8	14.4	14.6	14.5	13.1	12.9	7.2	7.2	5.4	7.0	9.3
Thr	6.9	7.0	8.3	7.0	11.1	9.1	6.9	8.5	3.7	4.8	7.9
Ser	6.7	6.8	6.8	6.3	11.2	9.9	6.9	5.4	22.4	20.0	17.3
Glu	9.0	9.5	8.3	10.0	6.4	8.5	3.7	2.4	13.5	13.1	8.4
Pro	7.2	6.1	6.1	6.1	5.2	6.4	7.3	7.4	1.5	5.1	6.1
Gly	7.8	6.8	7.4	6.3	8.5	7.7	8.3	8.0	25.8	20.0	8.7
Ala	9.0	7.1	6.8	7.0	6.5	7.6	18.0	19.4	8.0	9.1	14.4
Val	6.1	5.7	6.0	5.3	5.3	6.0	8.5	7.4	3.0	3.0	5.7
Met	2.3	2.7	3.0	2.7	1.5	1.5	traces	ND	0.4	0.5	0.3
Ile	3.4	3.6	3.9	4.4	5.0	4.2	2.0	2.5	1.5	2.3	1.8
Leu	7.4	7.7	7.3	7.8	9.9	8.4	2.4	2.2	2.2	2.7	8.5
Tyr	4.4	5.6	5.5	5.6	2.8	2.4	1.4	0.8	0.7	1.3	1.3
Phe	3.8	4.4	5.0	5.1	2.9	3.7	ND	ND	0.8	0.7	1.5
Trp	1.2	1.7	ND	ND	1.6	1.1	0.7	ND	ND	ND	ND
Half-cys	0.3	0.8	0.5	0.8	0.7	ND	ND	ND	ND	ND	ND

[a] POLASA and GREEN (1967).
[b] PETTERSSON (1971).
[c] SUNDQUIST et al. (1973a).
[d] PETTERSSON and HÖGLUND (1969).
[e] PRAGE and PETTERSSON (1971).
[f] EVERITT and PHILIPSON (1974).

ND = Not determined or not detected.

* The amino acid compositions are given in moles per 100 moles of amino acids.

could be detected in purified hexons (BISERTE et al., 1966; PETTERSSON et al., 1967; BOULANGER et al., 1969; LAVER, 1970) but later NEURATH et al. (1970a) showed that type 7 hexons could be labeled with ^{14}C-cystine and amino acid analysis under conditions which preserve half-cystine during hydrolysis has demonstrated that approximately 0.8 mol per cent of half-cystine is present in both ad2 and ad5 hexons (PETTERSSON, 1971) (Table 9). LAVER et al. (1967) discovered that all proteins in the outer capsid lack a free N-terminus and subsequently it was indeed found that hexons contain a blocked N-terminal amino acid (LAVER, 1970; PETTERSSON, 1971). The N-terminal peptide from ad2 hexon has been isolated. It can be labeled in vivo with ^{14}C-acetate and has the sequence Acetyl-Ala-Thr-Pro-Ser (JÖRNVALL et al., 1974a).

The polypeptide composition of the hexon was originally studied by MAIZEL et al. (1968b) with SDS polyacrylamide gel electrophoresis. They found that the hexon protein was a multimer of three identical polypeptides with a molecular weight of 120,000 daltons, a finding which has been confirmed by LAVER (1970). HORWITZ et al. (1970) estimated the molecular weight of the hexon polypeptide by sedimentation equilibrium in the presence of reducing agents and 6 M guanidine-HCl and could confirm the molecular weight obtained by SDS polyacrylamide gel electrophoresis. Estimation of the molecular weight by X-ray crystallography suggests values between 95—120,000 (CORNICK et al., 1971, 1973; FRANKLIN et al., 1971a). Biochemical studies including identification of the unique cysteine containing residues in the hexon (JÖRNVALL et al., 1974b) suggest that the hexon subunit contains 7 unique cysteines and a total of around 900 amino acid residues, which would indicate a molecular weight of 95—110,000. At present, it therefore appears well established that the molecular weight of the hexon polypeptide is in the range of 95—120,000 and thus that the hexon protein contains three polypeptide chains which seem to be identical.

3. Crystallization

The adenovirus hexon was the first animal virus protein to be crystallized. The original method used for ad5 hexon was described by PEREIRA et al. (1968). Hexons purified by DEAE chromaography and rate zonal centrifugation were dialyzed against 0.8 M KH_2PO_4. A precipitate rapidly formed which gradually converted into tetrahedral crystals. FRANKLIN et al. (1971a) used 0.5 M citrate buffer at pH 3.7—4.1 to generate bipyramidal crystals. Crystallographic studies of ad2 and ad5 hexon crystals (FRANKLIN et al., 1971a; CORNICK et al., 1971) revealed that both tetrahedral and bipyramidal crystals belong to the cubic system and have the space group $P2_13$. The length of the cubic cell was 149.9 Å and each unit cell contained 4 hexons. Since there are 12 asymmetric units in the cell, there has to be 3 crystallographic asymmetrical units per hexon. The cylinder-axis of the hexon is parallel to the axis of threefold symmetry. Therefore there are 3 structural units symmetrically arranged around the cylinder axis of the hexon which has also been confirmed by Patterson projection studies and by electron microscopy (FRANKLIN et al., 1971b; CROWTHER and FRANKLIN, 1972). One dyad perpendicular to the 3-fold axis and another dyad parallel to the 3-fold axis have been revealed (FRANKLIN et al., 1971b; LEIJONMARCK et al., 1974). The structural implication of these dyads is uncertain since the size of the hexon polypeptide infers that the hexon is a trimer. Homologous regions along the hexon polypeptide have not yet been ruled out (JÖRNVALL et al., 1974b) and could account for the additional symmetry elements.

4. Immunological Properties

Groupspecific determinants were early recognized in the adenovirus hexon and all adenoviruses except the avian and possibly some bovine (see section XIII) adenoviruses contain this antigenic specificity known as "α" (PEREIRA et al., 1963). Cross absorption studies and immunodiffusion (WILCOX and GINSBERG, 1963b; KÖHLER, 1965) revealed an additional typespecific determinant in the hexon referred to as "ε". A more complex arrangement of the antigenic determinants of the hexon was later suggested (NORRBY, 1969a and b; NORRBY and WADELL, 1969) since both type, group, intrasubgroup and intersubgroup specificities could be revealed by cross absorption and complement fixation studies. Hexons from ad 12 which elute at an aberrant position from DEAE-cellulose columns have only a few determinants in common with hexons from other serotypes (NORRBY and WADELL, 1969) (Table 10). Hexons from ad4, a serotype which is classified as a member of subgroup C (Table 7), share features with hexons of subgroup B and in particular with hexons from ad 16 (NORRBY and WADELL, 1969). Intermediate adenovirus strains, which contain hexon specificities from one parental strain and fiber specificities from another parental strain, have been described (NORRBY, 1969c). The intermediate strains ad 3—16 and ad 15—9 carry hexons which are antigenically similar but not identical to those of serotypes 3 and 15, respectively (NORRBY, 1969c).

Typespecific hexon determinants have been demonstrated on the surface of the virion by electron microscopy (NORRBY et al., 1969a). Groupspecific determinants on the other hand do not appear to be exposed since only homologous antihexon sera can form a corona-like pattern around virus particles as seen in the electron microscope (NORRBY et al., 1969a). Adenovirus from subgroup C seem to contain few determinants on the outside of the particle, since homologous hexon antibodies cannot aggregate the virions as revealed by electron microscopy (WADELL, 1972) and sucrose gradient centrifugation (PRAGE et al., 1970). Antibodies against the typespecific determinant of the hexon ("ε") interferes with hemagglutination of viruses from subgroups B and D probably because antibodies attached to the hexons in the peripentonal region prevent the attachment of the short fibers of group B and D viruses to the erythrocytes by steric hindrance (NORRBY, 1969b; NORRBY and WADELL, 1969). In contrast, subgroup C virions which have long fibers show less inhibition of hemagglutination by homologous hexon antibodies. The topography of the antigenic determinants on the ad 2 hexon polypeptide has been studied by limited proteolytic digestion (PETTERSSON, 1971). Trypsin removed 5—10 per cent of the polypeptides without changing the antigenic specificity. However, SDS polyacrylamide gel electrophoresis shows that the polypeptide is cleaved by this treatment in discrete fragments (unpublished), which are held together because of a rigid tertiary structure. Limited digestion with chymotrypsin, papain and subtilisin removed between 20—60 per cent of the hexon polypeptide and part of the groupspecific determinant "α" was eliminated. The typespecific determinant ("ε") was unchanged after proteolysis of native hexon, but could be inactivated by treatment of the hexon with maleic anhydride (PETTERSSON, 1971) which suggests that the "ε" determinant of ad2 hexons resides in a part of the hexon polypeptide which is unusually resistant to proteolysis.

B. The Fiber

Fibers from the excess pool in ad2-infected cells were originally purified by KÖHLER (1965) by methanol precipitation and DEAE chromatography and several authors in later reports have used repeated cycles of DEAE chromatography (RUSSELL et al., 1967a; HOLLINSHEAD et al., 1967; KJELLÉN and PEREIRA, 1968; MAIZEL et al., 1968b). Additional steps of purification are usually necessary to purify fibers to homogeneity. LEVINE and GINSBERG (1967) used DEAE cellulose and calcium phosphate chromatography and PETTERSSON et al. (1968) obtained homogeneous preparations of fiber by a 3-step purification procedure of DEAE-chromatography, isoelectric focusing and exclusion chromatography on 6 per cent agarose. Fibers are also made in excess in productive infection and as much as 25 $\mu g/10^6$ cells can be detected in extracts from infected cells by immunological techniques (EVERITT et al., 1971). Only 12.5 $\mu g/10^8$ cells could, however, be recovered (PETTERSSON et al., 1968) after purification either suggesting low recovery or a different structure of the fiber in the excess pool.

1. Morphology

The fiber is an antenna-like structure with a rod and a terminal knob (Fig. 19). The diameter of the rod is around 2 nm and about 4 nm for the knob. The length of the fiber varies for members of the different subgroups (Table 7). Subgroup B has the shortest fibers, 10 nm, and subgroup C the longest, 25—30 nm (NORRBY, 1969b). Ad4 an aberrant member of subgroup C has fibers which are 17—18 nm. Dimers of fibers have been detected in extracts from infected cells, but it is not known if these dimers contain an extra component which links the two units together (WADELL and NORRBY, 1969a; NORRBY et al., 1969b). Fibers from ad5 have been crystallized (MAUTNER and PEREIRA, 1971) but no detailed crystallographic study has yet been reported.

2. Physical-Chemical Properties

The sedimentation constant for fibers from ad2 and ad5 has been determined to be around 6S (LEVINE and GINSBERG, 1967; PETTERSSON et al., 1968; WADELL et al., 1969). Based on ultracentrifugation and electron microscopic studies the molecular weight for the fiber was estimated to be 60—80,000 (KÖHLER, 1965; VALENTINE and PEREIRA, 1965; HOLLINSHEAD et al., 1967; PETTERSSON et al., 1968). This seems to be an underestimate since calculations from gel filtration suggest a size 2—3 times larger (WADELL, 1970) and further determinations of the molecular weight by sedimentation equilibrium indicate a molecular weight of around 200,000 for the ad2 fiber (SUNDQUIST et al., 1973a). The polypeptide composition of the fiber has been studied by SDS polyacrylamide gel electrophoresis. One single class of polypeptides with a molecular weight of about 65,000 has been reported (PETTERSSON et al., 1968; MAIZEL et al., 1968b; SUNDQUIST et al., 1973a). These results would thus imply that there are 3 subunits in the ad2 fiber, but it remains to be demonstrated that all the subunits are identical, albeit they have equal molecular weight. The ad2 fiber has been found to be low in arginine and rich in hydroxyamino acid as compared to hexons (Table 9) (PETTERSSON et al., 1968). A half-cystine content of 0.7 per cent has been reported for the ad2 fiber (SUNDQUIST et al., 1973a), The fiber appears to be a glycoprotein; ISHIBASHI and MAIZEL (1974b) reported that ad2 fibers contain two residues of N-acetylglucosamine in each polypeptide chain. All the physical-chemical studies on the adenovirus fiber have so far been carried out with fibers from subgroup C viruses and no data are available on the structure of fibers from other subgroups.

3. Immunological Properties

A typespecific determinant ("γ") resides in the knob part of the fiber as revealed by electron microscopy (NORRBY et al., 1969a). The short fibers of serotypes belonging to subgroup B carry only the typespecific "γ" determinant. For serotypes with longer fibers a positive correlation has been found between length

Table 10. *Antigens Associated with the Major Structural Proteins*

Protein	Corresponding polypeptide (see Fig. 2)	Antigens[a]		
		Designation	Specificity	Remarks
Hexon	II	α	Group	Oriented towards the inside of the virion
		—	Inter- and intra-subgroup	
		ε	Type	Available at the surface of the virion from serotypes belonging to subgroups B and D
Penton base	III	β	Group	Carries toxin activity
		—	Inter- and intra-subgroup	
Fiber	IV	γ	Type	Reacts with HI-antibody
		—	Intersubgroup	Shared between members of subgroups C and D
		δ	Intrasubgroup	At the proximal part of the fiber only present in subgroups A, C and D (see Table 7)
Major core protein	VII		Subgroup Group and type?	Only ad 2 and ad 3 examined[b]

[a] Modified from WADELL (1970).
[b] Common as well as typespecific determinants have been revealed by immunodiffusion (PRAGE and PETTERSSON, 1971).

and antigenic complexity (Tables 7 and 10). The long fibers of virions from subgroups A, C and D have in the rod portion additional antigenic determinants (δ) which cross react between members of the same subgroup and some fibers also have inter-subgroup specific determinants (NORRBY, 1968, 1969b; PETTERSSON et al., 1968). The typespecific determinant can be revealed by hemagglutination inhibition or immunodiffusion (PEREIRA and FIGUEIREDO, 1962; VALENTINE and PEREIRA, 1965; NORRBY, 1966b; PETTERSSON et al., 1968). Electron microscopy has shown that this determinant attaches at least two IgG molecules (NORRBY et al., 1969a).

Antibodies against the "γ" specificity prevent attachment of virus to red cell receptors (PEREIRA and FIGUEIREDO, 1962) and also prevent the attachment of fibers to KB and HeLa cell receptors (PHILIPSON et al., 1968). Interaction of anti "γ" antibodies with virion causes aggregation of the virus particles (PRAGE et al., 1970), which makes it difficult to determine whether these antibodies prevent virus attachment in productive infection. The "δ" determinants are probably located near the junction between the fiber and the penton base since this determinant cannot be detected in the intact penton structure (PETTERSSON and HÖGLUND, 1969; WADELL and NORRBY, 1969).

The intermediate strain ad 15—9 carries fibers which are immunologically identical to ad 9 fibers, while the fibers from ad 3—16 are related but not identical to ad 16 fibers (NORRBY, 1969c).

C. The Penton

The penton has not been as thoroughly studied as the other structural proteins, primarily because it is difficult to purify it with good yields. Pentons from ad 2 and ad 5 have been purified by procedures similar to those used for hexons. DEAE chromatography alone (VALENTINE and PEREIRA, 1965; MAIZEL et al., 1968b) or combined with exclusion chromatography on 6 per cent agarose yields penton preparations which contain a few contaminants (PETTERSSON and HÖGLUND, 1969). Homogeneous preparations have been obtained after preparative polyacrylamide gel electrophoresis (PETTERSSON and HÖGLUND, 1969).

1. Morphology

The base of the penton exhibits a morphology simlar to that of isolated hexons and was originally described as a sphere with an average diameter of 8 nm (VALENTINE and PEREIRA, 1965). A conical morphology has been observed after negative staining with silicotungstate (LAVER et al., 1969; WADELL et al., 1969), while PETTERSSON and HÖGLUND (1969) observed globular structures sometimes showing a pentagonal outline after staining with uranyl acetate (Fig. 19). Central holes were observed in some penton bases but not as consistently as in hexon preparations. Free fibers and fibers attached to the penton base appear to have similar morphology. Aggregates of pentons have been isolated from cells infected with adenoviruses (NORRBY, 1966a; WADELL et al., 1969). Two types of aggregates have been recognized: dimers of pentons occur after infection with certain members of subgroups B and C. Some members of subgroups B and D as well as the aberrant serotype 4 produce aggregates with 12 pentons, with a perfect star-arrangement known as *dodecons* (Fig. 20). Although dodecons from some different serotypes contain 12 pentons of the same size, they sediment with widely different rates (50—100 S) (NORRBY, 1969 b). This may indicate that the pentons are oriented around a central core structure of unknown properties. The function or the structural role of these aggregates is not known.

2. Physical-Chemical Properties

The sedimentation coefficients for ad 2 and ad 5 pentons have been estimated to be 10.5—11 S (PETTERSSON and HÖGLUND, 1969; WADELL et al., 1969). PETTERSSON and HÖGLUND (1969) estimated a value of 400,000 daltons for ad 2 pentons by sedimentation diffusion and WADELL (1970) suggested 485,000—505,000 for ad 5 pentons by gel chromatography. The molecular weight of ad 5 pentons was estimated to be 280,000 by electron microscopy (VALENTINE and PEREIRA, 1965), which probably is an underestimate due to shrinkage during fixation. The bonds between the base and the fiber are noncovalent and can be disrupted by 2.5 M guanidine-HCl (NORRBY and SKAARET, 1967), 33 per cent formamide (NEURATH et al., 1968) or 8 per cent pyridine (PETTERSSON and HÖGLUND, 1969). After dissociation of the penton structure the base can be purified by electrophoresis with retention of its antigenicity and ability to induce characteristic cytopathic changes (see section XII: C: 2) in monolayers of HeLa or KB cells (PETTERSSON and HÖGLUND, 1969). In most cases free vertex capsomeres do not seem to be present in a large excess in infected cells. NORRBY and ANKERST (1969) isolated free penton bases from cells infected with ad 12. WINTERS et al. (1970) have subsequently reported that free bases from cells infected with ad 5 can be isolated with preparative polyacrylamide gel electrophoresis. It is not clear whether penton bases ever exist free *in vivo* or if they are formed during purification because of degradation of pentons.

Two polypeptides have been detected in ad 2 pentons by SDS-polyacrylamide gel electrophoresis (MAIZEL et al., 1968 b). One corresponds to the polypeptide of the fiber and the second has a molecular weight of around 70,000. No information is available on the number of polypeptides in the penton base. The amino acid composition of pentons from ad 2 has been determined and pentons resemble hexons although significant differences are observed, in particular with regard to tyrosine and the hydroxyamino acids (Table 9).

The penton base is sensitive to proteolytic degradation by trypsin. At high enzyme concentration the base is destroyed and fibers and low molecular weight products are obtained. Low concentrations of trypsin abolish the cytopathogenic effect but the antigenic properties are preserved (PETTERSSON and HÖGLUND, 1969; WADELL and NORRBY, 1969 b).

3. Immunological Properties

The penton base appears to carry a weak antigenic determinant known as "β" which is common to pentons from all adenovirus serotypes. In addition, inter- and intrasubgroup specific determinants have been revealed (WADELL and

NORRBY, 1969b, Table 10). All these determinants have been demonstrated by hemagglutination enhancement while immunodiffusion mainly recognizes the intrasubgroup specific antigen. Pentons from ad 12 appear to carry only the group reactivity and no additional determinants have been detected (WADELL and NORRBY, 1969).

D. The Major Core Protein

The major core protein from ad 2 and ad 3 has been extensively purified (LAVER 1970; PRAGE and PETTERSSON, 1971; EVERITT et al., 1973). A precursor protein to the major core protein has recently been described (ANDERSON et al., 1973). The precursor contains 5 methionine-containing tryptic peptides whereas the final product lacks one of these peptides. The precursor polypeptide may be extracted by acid urea from nuclei of infected cells late in infection and it seems to be synthesized in excess (EVERITT and PHILIPSON, 1974). Purification of the major core protein has been achieved only with disrupted virions as starting material. LAVER (1970) disrupted ad 2 virions with SDS and fractionated the viral proteins into three classes by ethanol precipitation. Further purification was achieved by SDS polyacrylamide electrophoresis and a homogeneous preparation of the major core polypeptide was obtained. PRAGE et al. (1970) disrupted purified ad 2 and ad 3 virions by freezing and thawing which released hexons and pentons as separate units and the adenovirus DNA was obtained as a precipitate with the associated core proteins. The core proteins were extracted with dilute acids and preparative polyacrylamide electrophoresis at pH 4.6 without SDS was used for the final purification (PRAGE and PETTERSSON, 1971). The major core protein corresponds to polypeptide VII in the SDS polyacrylamide gel system of MAIZEL and coworkers (MAIZEL et al., 1968; EVERITT et al., 1973; ANDERSON et al., 1973) and component VIII in the corresponding system of LAVER (1970). The molecular weight of the major core protein has been estimated to be 18,000 by equilibrium centrifugation (PRAGE and PETTERSSON, 1971), and SDS polyacrylamide gel electrophoresis (MAIZEL, 1971; EVERITT et al., 1973; ANDERSON et al., 1973). The protein contains one single polypeptide chain with two residues of tyrosine, one residue of tryptophan and with alanine as the N-terminal amino acid. The amino acid analysis (Table 9) shows that the protein is basic with about 23 per cent arginine. It resembles the arginine-rich histones with regard to the arginine and the alanine content (19 per cent), but differs from these because it contains tryptophan and lower amounts of lysine. The arginine residues are probably evenly distributed along the polypeptide chain (LAVER, 1970) and it has been estimated that half of the phosphate groups on the adenovirus DNA could be neutralized by arginine residues of the core protein (PRAGE and PETTERSSON, 1971). However, no evidence has been obtained that this is the function of the core protein *in vivo*. The major core protein like other basic proteins is weakly antigenic and special precautions must be taken to reveal precipitation lines by immunodiffusion (PRAGE and PETTERSSON, 1971). Core proteins from ad 2 and ad 3, which belong to different subgroups, share some immunological properties, but differences in antigenic determinants are also apparent (Table 10).

E. Low Molecular Weight Proteins of the Virion

High resolving SDS polyacrylamide gel electrophoresis of adenovirus (MAIZEL, 1971; EVERITT et al., 1973; ANDERSON et al., 1973) reveals at least 10 polypeptide bands (see section III). An investigation of the low molecular weight proteins of the virion has been carried out by EVERITT and PHILIPSON (1974). Virion proteins were extracted in a two-step procedure with urea at two different pH values. The basic proteins V, VI, VII and VIII were extracted at pH 3.1 and separated by preparative polyacrylamide gel electrophoresis in acid urea. Each individual protein could be further purified to homogeneity by one or two steps of exclusion or ion-exchange chromatography in acid urea. Protein VI, which appears to have a molecular weight of around 23,500 in SDS, probably exists as a dimer under native conditions since gelfiltration without SDS gives a molecular weight of about 50,000 (EVERITT and PHILIPSON, 1974). The amino acid compositions of protein VI and VIII show distinct differences when compared to the compositions of hexons, pentons and fibers (Table 9).

Protein IX was extracted by alkaline urea and purified to homogeneity by ion-exchange chromatography on QAE Sephadex. It has a molecular weight of around 12,000 and the amino acid composition shows a large number of acidic residues (Tables 8 and 9). The three low molecular weight proteins VI, VIII and IX are antigenically distinct and show no antigenic relatedness to the major capsid proteins. However, it is difficult to completely exclude that these proteins are cleavage products from the major capsid proteins until their primary structures have been determined.

XII. Physiological Effects of the Structural Proteins

Since adenoviruses contain more polypeptides than many other viruses, it is conceivable that some virion proteins have specific functions in addition to their structural role. It may therefore be rewarding to study the role of the structural proteins in viral DNA replication, transcription and translation. A ligase or proteins controlling DNA polymerase activity may be carried by the virion since integration of viral DNA appears to be independent of cellular DNA and protein synthesis (DOERFLER, 1970), but preexisting enzymes may also be utilized. The presence of positive control factors for the cellular DNA-dependent RNA polymerases may explain why early RNA synthesis in productive infection is independent of protein synthesis (PARSONS and GREEN, 1971). So far, only one enzymatic activity has been ascribed to a structural protein but considerable information has accumulated on the role of the structural proteins in virus-cell interaction, in neutralization and also with regard to the effect of the structural proteins on enzymes involved in macromolecular synthesis. These aspects will be reviewed in this section.

A. Hemagglutination

Hemagglutination by adenoviruses was first demonstrated by ROSEN in 1958. All human types have subsequently been shown to display hemagglutinating capacity (ROSEN, 1960; ROSEN et al., 1962; BAUER and WIGAND, 1963; SCHMIDT *et al.*, 1965; NORRBY, 1969b). The interaction of viruses with erythrocytes has been considered to be a model system for studies of virus-cell interaction and it appears established that the vertex region of the virion is involved in the interaction between viruses and red cells. When the structural units of the virion were isolated and morphologically defined, it was shown that the typespecific determinant ("γ") of the fiber interacts with the red cell surface (PEREIRA and FIGUEIREDO, 1962; VALENTINE and PEREIRA, 1965; NORRBY, 1966b; PETTERSSON *et al.*, 1968; NORRBY *et al.*, 1969a). Specific receptors seem to be present on the erythrocyte membrane and NEURATH *et al.* (1969) have solubilized and partially purified an adenovirus receptor from monkey red cells.

Based on their ability to agglutinate different red cells, ROSEN (1960) proposed a subgroup classification of the human adenoviruses. Subgroup I viruses agglutinate monkey erythrocytes, preferably from grivet and rhesus species, with a complete pattern[1]. Subgroup II viruses agglutinate rat erythrocytes with a complete pattern, and subgroup III viruses agglutinate rat cells with a partial pattern. Subgroups I, II and III correspond to subgroups B, D and C as defined by oncogenicity (HUEBNER, 1967; Table 7). Viruses in subgroup IV which includes the highly oncogenic subgroup A viruses were originally thought to lack hemagglutination capacity but have subsequently been shown to agglutinate rat cells with a partial pattern (SCHMIDT *et al.*, 1965; NORRBY, 1969b). The partial agglutination is due to a competition between mono- and multivalent virus components for the receptors on the erythrocyte (WADELL, 1969). Further subdivisions have been suggested, based on the species origin of the erythrocytes agglutinated by different virus types within one subgroup (for reviews see NORRBY, 1968, 1969b).

Since the intact virion carries several vertex projections they can establish a bridge between the erythrocytes, and give a complete hemagglutination pattern. Multimers of pentons and the fibers like dimers of pentons and fibers as well as dodecons can, in the same way, give rise to hemagglutination (NORRBY, 1966a; WADELL *et al.*, 1969; WADELL and NORRBY, 1969a; NORRBY *et al.*, 1969b). Monomers of penton or fiber can, on the other hand, only establish a monovalent link with the cell and agglutination can only be detected when antibodies are used to bridge the erythrocyte-associated structural units (Fig. 20) (PEREIRA and FIGUEIREDO, 1962; NORRBY, 1966b; NORRBY, 1969b). Since the typespecific determinant of the fiber is responsible for erythrocyte attachment, heterotypic antibodies must be used to demonstrate agglutination by penton and fiber monomers (PEREIRA and FIGUEIREDO, 1962; NORRBY and SKAARET, 1967). The penton base carries groupspecificity and any heterotypic antiserum will thus make pentons into multivalent hemagglutinins (Fig. 20). This method of revealing adenovirus hemagglutination by heterotypic antisera has been called the *hemagglutination enhancement* (HE) reaction (ROSEN, 1960) (Fig. 20). Similarly, fibers from subgroups II and III which carry inter- and intrasubgroup specific determinants are made into divalent agglutinins by heterotypic antibodies. The short

---

[1] A complete pattern refers to a hemagglutination pattern where the entire bottom of the tube is covered by an equally dense lattice. A partial pattern refers to a pattern where a fraction of the erythrocytes has sedimented without agglutination and form a ring in a less dense lattice.

fibers of subgroup I viruses cannot agglutinate since they only contain typespecific determinants (NORRBY and SKAARET, 1967). Antibodies directed toward the γ-specificity of the fiber can be assayed by *hemagglutination inhibition* (HI) tests (Fig. 20). Specific antibody consumption tests have been devised to assay for the

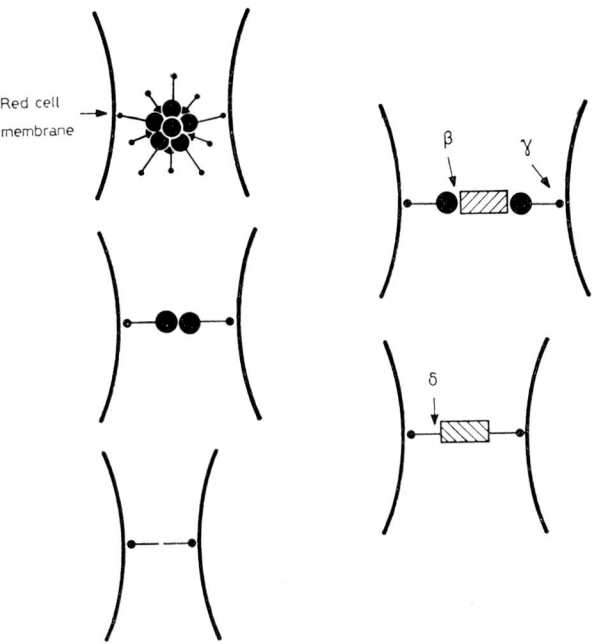

Fig. 20. A diagrammatic representation of the structural units of adenoviruses which can cause hemagglutination.
Dodecons (top, left) consist of 12 pentons in a star-like arrangement. Dimers of pentons (middle, left) and dimers of fibers (bottom, left) can cause hemagglutination since they are divalent and can therefore aggregate red cells. Monovalent units like isolated pentons and fibers (top-right and bottom-right, respectively) can only hemagglutinate after being linked with heterotypic antibodies. Heterotypic antibodies against the penton base are required to reveal *hemagglutination enhancement* for pentons and heterotypic antibodies against fibers are necessary to reveal hemagglutination with fibers. The antigenic determinants of the fiber and the penton base are indicated in the Figure as defined in Table 10. Modified from NORRBY (1966)

group, inter- and intrasubgroup specific determinants on penton bases and fibers. Penton base may thus be assayed by a *hemagglutination enhancement consumption* (HEC) test (NORRBY and SKAARET, 1967). Free fibers from subgroup I which lack subgroup specificity, may be assayed by a *hemagglutination inhibition consumption* (HIC) test (NORRBY and SKAARET, 1967). With the aid of these different techniques NORRBY and coworkers (for reviews see NORRBY, 1968, 1969b; WADELL, 1970) have elucidated the immunological relationship between the capsid components from different adenovirus serotypes.

B. Neutralization of Adenoviruses

Adenoviruses carry a mosaic of antigenic specificities on their surface. Since by definition all types are distinct by neutralization, typespecific antigens should induce neutralizing antibodies against these viruses. Both the fiber and the hexon show distinct typespecific antigenic determinants, the "γ" and "ε", respectively, and both have been claimed to induce neutralizing antibody (WILCOX and GINSBERG, 1963b; KASEL et al., 1966). The method to determine neutralization is of importance with adenoviruses, since these viruses have a slow multiplication cycle. Assays based on an all or none response, such as scoring for cytopathic effect, are probably the most sensitive since extremely low virus concentrations can be used. On the other hand, they may be less specific since they allow for secondary reactions between virus and antibody during the long incubation periods (approximately 20 days). Plaque reduction or fluorescent focus inhibition (PHILIPSON et al., 1968) measure virus-antibody interaction in a more direct way. These two assays differ significantly with regard to incubation time: 8—10 days for the plaque assay versus 48 hours for the fluorescent focus assay. Reversible virus-antibody interactions and delay in virus eclipse may be overlooked in the plaque assay which depends on several reproductive cycles. Since the structural proteins of adenoviruses are available in a highly purified state, it should be possible to define the structural protein involved in virus neutralization, assuming that a single polypeptide is responsible for eliciting neutralizing antibodies.

Adenovirus neutralization is a matter of considerable controversy. Several early studies showed that antisera against partially purified hexons induced neutralizing antibodies (WILCOX and GINSBERG, 1963b; KASEL et al., 1966; KJELLÉN and PEREIRA, 1968). Hexons purified by crystallization were also found to induce neutralizing antibodies (PEREIRA and LAVER, 1970). In contrast, PETTERSSON et al. (1967) found that antisera against electrophoretically purified hexons did not neutralize virus infectivity although such hexons contained both group- and typespecific antigenic determinants (PETTERSSON, 1971). The controversy has not yet been settled but some recent reports have given additional information. Antisera against purified hexons from subgroups B and D viruses seem to contain potent neutralizing antibodies (for a review see WADELL, 1972). Electron microscopy of complexes between subgroup C virions and homologous hexon antisera show that the hexons of these viruses contain no or very few exposed antigenic determinants (WADELL, 1972). Hexons from cells infected with ad2 have been separated into two classes by electrophoresis (PETTERSSON, 1971) and ion-exchange chromatography or isoelectric focusing (WADELL, 1972). Antisera against the major class of hexons ("fast migrating hexons") do not neutralize virus infectivity, whereas the minor class was found to induce neutralizing antibodies. It has not yet been possible to show any difference in the polypeptide composition of these two classes of hexons.

Antisera against subgroup C fibers do not contain neutralizing activity as measured with the plaque assay (PETTERSSON et al., 1968; KJELLÉN and PEREIRA, 1968). On the other hand, when neutralization is scored by the fluorescent focus assay, fiber antisera were found to contain high titers of neutralizing activity. There is yet no experimental data available to explain this difference but it seems likely that fiber antibodies cause a reversible inhibition or delay of the infection. This effect is presumably not scored by the plaque techniques, since the virus goes through several reproductive cycles in this assay. Antisera against short fibers from subgroup B virus also seem to have neutralizing activity when measured with assays which allow for several infectious cycles (NORRBY, 1969b).

Antisera prepared against disrupted ad2 virions are considerably more efficient in neutralization than antisera against any of the purified major capsid components. Recently, additional components have been identified in the adenovirus capsid (see section III on composition of the virion). Some of these are now available in a purified form and it seems necessary to evaluate the neutralizing effect of antisera against these components before neutralization of adenovirus infectivity is completely understood. Preliminary results indicate that polypeptide IIIa from ad2 may induce neutralizing antisera (EVERITT and PHILIPSON, unpublished).

C. Possible Functions of Individual Adenovirus Proteins

1. Hexon

It is not yet known if hexons play any physiological role during adenovirus infection. The majority of the hexons remain associated with the infectious virus when it penetrates the plasma membrane in productive infection (LONBERG-HOLM and PHILIPSON, 1969), but it appears that the hexons are shed in the cytoplasm before the viral DNA enters the nucleus.

It has been shown that hexons can bind to DNA and inhibit DNA polymerase or DNA-dependent RNA polymerase *in vitro* (LEVINE and GINSBERG, 1968). The significance of this inhibition during infection is unclear, since ts-mutants which are temperature-sensitive for synthesis of structural proteins and for viral DNA synthesis still show inhibition of host cell DNA synthesis (WILKIE et al., 1973).

2. Penton

In productive adenovirus infection characteristic cytopathic changes can be detected. The cells have a granular cytoplasm, round up and detach, but the individual cell does not rupture or expel the cytoplasm until very late after infection. PEREIRA and KELLEY (1957a) established that 2 types of cytopathic effects could be discerned. The first is apparent within 8—24 hours after infection with high concentrations of crude virus preparations, and the second which is accompanied with virus production can be seen after 7—20 days at low multiplicities of infection. Later, EVERETT and GINSBERG (1958), PEREIRA (1958) and ROWE et al. (1958) established that a protein from infected cells, which sedimented slower than the virus, could induce the early cytopathic changes. When the structural components of the adenovirion were identified, this early cytopathic factor (toxin) was found to be associated with the pentons (VALENTINE and PEREIRA, 1965). Penton concentrations of about 0.1 µg/10^6 cells are required for a clearcut effect. The changes in cell morphology do not impair macromolecular synthesis and the effect is fully reversible after washing with fresh medium (PEREIRA, 1958; ROWE et al., 1958). The molecular mechanism for the cytopathic effect is unknown. The finding that lipid metabolism of the host cell is enhanced early after adenovirus infection (McIntosh et al., 1971) and that the penton base alone can induce an increase in lipid synthesis may suggest a route for further experiments.

Pentons from certain serotypes have been found to lack toxin activity (Wadell, 1970). The cytopathic effect of the penton is particularly susceptible to proteolytic degradation by trypsin and, at low concentrations of the enzyme, it is possible to destroy this effect without changing the antigenic properties of the penton base (Pettersson and Höglund, 1969; Wadell and Norrby, 1969b). The cytopathic effect of pentons from subgroup C is neutralized by anti-penton sera but not by sera against fibers (Pettersson and Höglund, 1969; Wadell and Norrby, 1969b), which suggests that the penton base alone carries the cytopathic effect. In accord with this, it has been found that isolated penton bases are able to induce early cytopathic changes (Pettersson and Höglund, 1969; Winters et al., 1970).

An endonuclease activity has been found to be associated with the ad2 pentons (Burlingham et al., 1971). Burlingham and Doerfler (1972) detected an endonuclease in cells infected with ad2 and ad12. This endonuclease was also found in preparations of purified virions. When highly purified structural components were tested for activity, only pentons were found to have an effect (Burlingham et al., 1971). The hexon and the fiber were inactive, even at high concentrations, and they did not compete with the penton-associated endonuclease *in vitro*. The endonuclease appears to be specific for DNA and cleaves double stranded DNA 20 times faster than denatured DNA with a preference for GC-rich regions. Recent results suggest that the adenovirus endonuclease can be dissociated from pentons by treatment with high salt concentration and it has been extensively purified by a multistep procedure involving ion-exchange and exclusion chromatography (Doerfler, personal communication). It has not yet been established that this endonuclease is a virus-coded component, but a similar activity is not detectable in uninfected cells (Doerfler, personal communication).

3. Fiber

Since the fiber is responsible for attachment of adenoviruses to erythrocytes, it is assumed that the fiber recognizes receptors on the plasma membrane in productive infection and thus mediates early virus-cell contact. Evidence to substantiate this assumption is available from studies on adenovirus infection in KB and HeLa cells (Levine and Ginsberg, 1967; Philipson et al., 1968). Purified fiber preparations in quantities 10- to 100-fold higher than the number of virus receptors (around 10^4/cell) can inhibit virus attachment to the cell surface (Philipson et al., 1968). With the aid of fiber preparations which were labeled *in vitro* with iodine it could be demonstrated that the fiber indeed attaches to cells and the attachment was prevented by cold fiber or fiber antibodies (Philipson et al., 1968). After penetration of the plasma membrane the virus seems to lose the fiber together with the penton base and some hexons (Sussenbach, 1967). Experiments with ^{125}I-labeled virions have shown that intracellular virus lacks the fiber and other polypeptides associated with the peripentonal region (EVERITT, personal communication).

High concentrations of fiber can also interfere with replication of adenovirus as well as with polio and vaccinia virus, and the fiber seems to correspond to the *adenovirus interfering principle* first described by PEREIRA (1960). Like hexons, fibers bind to cellular as well as to viral DNA at low ionic strength. This binding inhibits the activity of DNA and RNA polymerase. It has been claimed that purified fiber can inhibit macromolecular synthesis in uninfected cells after a latent period of 24—28 hours (LEVINE and GINSBERG, 1967, 1968). DNA, RNA and protein synthesis was reduced when ad5 fiber was applied to cell cultures at a concentration of 50 µg/10^6 cells. This effect may, however, have little relevance for inhibition of host cell macromolecular synthesis *in vivo* since inhibition of host macromolecular synthesis in the productive cycle occurs prior to detectable synthesis of the fiber protein (RUSSELL et al., 1967a). Also mutants which are temperature-sensitive for synthesis of fiber still inhibit host cell DNA synthesis at the nonpermissive temperatures (WILKIE et al., 1973).

No enzymatic effect has so far been associated with the fiber. Purified fibers from ad1 and ad2 carry the erythrocyte receptor modifying effect (ERM) which interferes with receptors on human erythrocytes for certain subgroup D members (KASEL et al., 1961; KASEL and HUBER, 1964). Later studies indicated, however, that this effect is due to a steric block of the erythrocyte receptors by fibers rather than to an enzymatic modification of the receptor (WADELL, 1969).

4. Virus-Induced Non-Structural Proteins

Infected cell extracts contain several polypeptides which are not present in uninfected cells and which do not constitute structural proteins of the virion (RUSSELL and SKEHEL, 1972; ANDERSON et al., 1973; WALTER and MAIZEL, 1974). The proteins containing these polypeptides have not yet been purified but indirect evidence points to a function for some of these polypeptides. Two polypeptides with molecular weights of 71,000 and 48,000 have been purified from infected cells by DNA-cellulose chromatography (VAN DER VLIET and LEVINE, 1973). They have a strong affinity for single stranded DNA and are most likely involved in the replication of adenovirus DNA. Their properties have been described in a previous section (section V: F). A polypeptide with a molecular weight of 100 K is induced during adenovirus infection and is associated with messenger ribonucleoprotein particles from infected cells (LINDBERG and SUNDQUIST, 1974). This polypeptide might be necessary for proper initiation of translation of virus specific messenger RNA.

XIII. Classification and Nomenclature of Adenoviruses

A. Adenoviruses from Different Species

More than 80 different adenovirus types have been isolated from a variety of animal species and Table 11 lists the types which have been described. All adenoviruses seem to have a similar morphology which was outlined in the section on the architecture of the virion (section II). Differences have been noticed with regard to the length of fibers from different types and one member of the avian group "CELO" appears to contain pentons which have two fibers of different

Table 11. *Adenoviruses Isolated from Different Species*[a]

Natural host	Serological types	References
Human	28 accepted 31 recognized	BÉLÁDI (1972)
Simian	23	HILLIS and GOODMAN (1969)
Bovine	8	MOHANTY (1971) BARTHA, personal communication
Equine	1	STUDDERT et al. (1974)
Ovine	1	McFERRAN et al. (1969)
Canine	2	MARUSYK et al. (1970)
Murine	2	HARTLEY and ROWE (1960) REEVES et al. (1967)
Porcine	4	BIBRACK (1969)
Avian	8	McFERRAN et al. (1972)

[a] Modified from NORRBY (1971).

length (42.5 nm and 8.5 nm) (LAVER et al., 1971). Although the criteria used for differentiating adenoviruses into serotypes have been based primarily on specific neutralization of infectivity with antisera, the question is still raised as to whether neutralization or hemagglutination inhibition tests (NORRBY, 1971) should be the parameter used for differentiation of adenoviruses. As discussed in detail in a previous section (section XII:B), it has been suggested that the typespecific antigen involved in virus neutralization resides in the hexon (KJELLÉN and PEREIRA, 1968). Hexon antibodies seem to efficiently neutralize infectivity of viruses from subgroups B and D, whereas antisera prepared against highly purified hexons from subgroup C fail to neutralize infectivity efficiently (PETTERSSON et al., 1967; WADELL, 1972). Recent studies suggest that polypeptide IIIa (Fig. 2) is of importance for induction of antibodies which neutralize ad 2 infectivity (EVERITT and PHILIPSON, in preparation). Antibodies which inhibit hemagglutination are induced by an antigen which resides in the fiber. It has been proposed that serotypes should be differentiated by neutralization tests and isolates distinguishable only in hemagglutination inhibition tests should be considered as subtypes of the serotype identified with the former technique (NORRBY, 1969b). More detailed reviews of the classification of adenoviruses have recently been published (NORRBY, 1971; ANDREWES and PEREIRA, 1972).

Although only 31 human serotypes have been officially recognized (BELADI, 1972), virus neutralization tests distinguish 33 serotypes of human adenoviruses (BLACKLOW et al., 1969). Cross reactions have been observed among a few different serotypes (RAFAJKO, 1964; WIGAND et al., 1965) and the cross reaction between serotype 29 and serotype 15 is of such a magnitude that ad 29 should be regarded as a subtype of ad 15 (STEVENS et al., 1967). Intermediate strains of human adenoviruses with hexon and fiber units which are derived from two different parental serotypes have been described (NORRBY, 1969c). These strains include the intermediate strain ad 15—9 and the *San Carlos agent*, an intermediate strain of ad 3 and ad 16. Phenotypically mixed particles can be generated by mixed infections with different human adenovirus strains (NORRBY and GOLLMAR, 1971).

The simian adenoviruses have been divided into 2 groups (HILLIS and GOODMAN, 1969); those from monkeys and those from chimpanzees. Seven serotypes from chimpanzees have been identified by neutralization and hemagglutination inhibition tests. Several serotypes from chimpanzees display characteristics which indicate that they are related but not identical to human serotypes ad 14, ad 18, ad 5 and ad 2. RAPOZA (1967) listed eighteen monkey serotypes, but six of these were related pairwise in neutralization tests, so only fifteen seem to be unique. Findings by KIM et al. (1967) suggest the addition of one more serotype in the monkey group giving a total number of sixteen monkey serotypes and thus a total of twenty-three simian adenovirus serotypes.

At present, eight serotypes of bovine adenoviruses are known (MOHANTY, 1971; BARTHA, personal communication). Two additional serotypes have been reported but appear to crossreact with bovine serotypes 4 and 6. Types 1, 2, and 3 share the common groupspecific antigen ("α") which can be detected in human and most other adenoviruses. Like the avian adenoviruses, bovine serotypes 4 to 8 lack this common adenovirus antigen (BARTHA, personal communication). Another "groupspecific" antigen can be detected in bovine types 4 to 8 which does not crossreact with human or avian adenoviruses.

Ovine adenoviruses were identified only recently and no comparative analysis of the eight isolates by McFERRAN et al. (1969) has been reported.

From the porcine group four distinct serotypes have been isolated by several investigators and these viruses all contain the groupspecific adenoviruses antigen and no cross reaction could be detected with adenoviruses from other hosts by virus neutralization (CLARKE et al., 1967; KASZA, 1966).

The canine laryngotracheitis virus appears to be distinct from the infectious canine hepatitis virus in hemagglutination inhibition and virus neutralization tests, although the latter test reveals some cross reactivity (SWANGO et al., 1969; MARUSYK et al., 1970). It has been suggested that the ICH agent is antigenically related to human ad 8, since humans have been found to carry antibodies against this virus (SMITH et al., 1970).

The first murine adenovirus was isolated by HARTLEY and ROWE (1960) during attempts to establish Friend leukemia virus in cell cultures. VAN DER VEEN and MES (1974) have found that this strain is distinct from the murine isolate described by HASHIMOTO et al. (1966) and REEVES et al. (1967). Other murine strains have been isolated by MISSAL (1969) but their relationship to other isolates is unknown.

KAWAMURA et al. (1964) described eight different avian serotypes including Gallus adeno-like (GAL) and CELO (Chicken Embryo Lethal Orphan) viruses, and more recent data by McFERRAN et al. (1972) seem to confirm that there are at least eight distinct serotypes within the avian group. It is noteworthy that all the avian adenovirus serotypes lack the adenovirus groupspecific antigen but instead share a common antigen which can be revealed by immunodiffusion and also, but less efficiently, by complement fixation (KAWAMURA et al., 1964).

Viruses morphologically identical to adenoviruses have also been isolated from horses (TODD, 1969) and serological tests have shown that equine sera often contain antibodies which react with the common antigen of mammalian adenoviruses (DARBYSHIRE and PEREIRA, 1964; TIMONEY, 1971). STUDDERT et al. (1974) have compared six isolates of equine adenoviruses which originated from the United States, Germany and Australia and they were all closely related.

In addition, WILNER (1969) reports of an adenovirus-like agent from oppossum.

B. Subgroup Classification

Several different parameters can be used to classify the adenoviruses from animal species into subgroups. Biological characteristics such as host range, capacity to transform cells *in vitro* and the ability to induce tumors in experimental animals can be used. Since the capsid subunits from different adenoviruses can be purified and characterized, it is also possible to base a subgroup classification on their morphological and immunological properties. Ultimately it is desirable to base the subgroup classification on sequence differences between genomes from different serotypes as suggested by LWOFF et al. (1962) for animal viruses in general. Two different subgroup classifications of human adenoviruses have been proposed; one is based on the hemagglutination properties (ROSEN, 1960) and the other is based on the oncogenicity for newborn hamsters (HUEBNER, 1967). The first classification separates the human adenoviruses into four groups on the basis of their ability to agglutinate monkey and rat erythrocytes as described in detail in section XII: A. Group I causes complete agglutination of monkey erythrocytes. Group II causes complete agglutination of rat erythrocytes. Group III causes incomplete agglutination of rat erythrocytes, and group IV was originally thought not to cause detectable hemagglutination. Subsequent to this classification it was reported (SCHMIDT et al., 1965; NORRBY and ANKERST, 1969) that ad 12, ad 18 and ad 31, members of group IV, could agglutinate rat erythrocytes with an incomplete pattern, in a similar way as group III viruses. Members of the same subgroups have been shown to have fibers of a characteristic length, *i.e.* subgroup I has fibers which are about 10 nm, subgroup II 12—13 nm, subgroups III and IV 23—31 nm (Table 7; NORRBY, 1969b). Adenovirus types 20, 25 and 28, which were originally placed in hemagglutination group I (ROSEN, 1960), probably belong in group II, both with respect to the pattern of hemagglutination and also with regard to the length of the fibers (WIGAND, 1970). Serotype 4 which belongs to subgroup III has some anomalous biological properties, and has fibers which are intermediate in length to those of group II and III viruses. In several respects, ad 4 shares many features with serotype 16 of subgroup II (for a review see NORRBY, 1968). Thus, the original classification based on hemagglutination pattern has created some confusion about the classification of the oncogenic ad 12, ad 18 and ad 31 and also of other serotypes. For human adenoviruses the classification based on oncogenicity (HUEBNER, 1967) appears to better correlate with the structure of the DNA. First the GC content of the viral DNA differs between the groups A, B, and C so that the highly oncogenic viruses (subgroup A) have 48—49 per cent GC, the weakly oncogenic (subgroup B) 50—52 per cent GC, the non-oncogenic viruses (subgroups C and D) contain 57—59 per cent GC (reviewed by GREEN, 1970). Non-oncogenic serotypes in subgroups C and D are distinguished because they induce serologically different T-antigens (MCALLISTER et al., 1969b). Secondly, the classification of human adenoviruses in subgroups A—D correlates with sequence homology as detected by hybridization. Extensive homology between members of the same oncogenic subgroup has been demonstrated by filter hybridization (LACY and GREEN, 1964, 1965, 1967; Table 2) and by heteroduplex mapping (GARON et al., 1973, see also section IV). It is therefore suggested that the classification of human adenoviruses into subgroups A—D which appears to correlate with sequence homology, should be accepted. As evident from Table 7, this classification correlates well with the classification based on hemagglutination, provided that the serotypes ad 12, ad 18 and ad 31 are considered as a separate hemagglutination group, as originally suggested by ROSEN (1960). It should be emphasized that not all types have been studied with regard to both DNA homology and biological parameters and the proposed classification in Table 7 is therefore tentative. In particular, group D viruses have not been analyzed in detail.

The antigenic relationship between capsid antigens from different adenovirus serotypes has been extensively investigated by NORRBY and coworkers (see reviews by NORRBY, 1968, 1969b, 1971, and WADELL, 1970). It is possible that the group and type specificities of the hexon as well as the group and subgroup specificities of the penton and certain fibers could be used for more refined subgroup classification of the human adenoviruses and also help to elucidate their evolution from a possible common ancestor.

The subgroup classification for adenoviruses isolated from animal species is not yet elaborate and has in most cases been based on the hemagglutination properties of the isolated strains. RAPOZA (1967) divided the monkey adenoviruses into four subgroups according to hemagglutination behaviour and adenoviruses from chimpanzees were divided into three hemagglutination subgroups (HILLIS and GOODMAN, 1969). Among the simian adenoviruses it is apparent that the oncogenic serotypes are not associated with a particular hemagglutination group since the oncogenic adenoviruses SV 20, SV 23, SV 34, SV 37, SV 38 and SA 7 are distributed among the three different subgroups. Bovine adenoviruses have been divided into two subgroups. Subgroup 1 includes serotypes 1, 2 and 3 which contain the common adenovirus groupspecific antigen and they can be propagated on most bovine cell cultures. Subgroup 2 viruses (serotypes 4 to 8) have a different group antigen and have so far only been grown on bovine testicle cells (BARTHA, personal communication). Hemagglutination has also been observed for some members of ovine, porcine and canine adenoviruses, but no hemagglutinin has been isolated from the murine and the avian groups.

Although all adenoviruses are similar in their gross architecture they show a wide antigenic variation. It might be rewarding to try to demonstrate variable and constant portions in the polypeptides of the structural proteins to understand the diversification. Only the major capsid polypeptides of the virion has yet been immunologically compared between different adenoviruses. The contribution of other polypeptides than those of the hexon, penton and fiber to neutralization and other biological properties may therefore also be rewarding.

XIV. Adenovirus Infection in Humans and Animals

A. Human Adenovirus Infections

1. Pathogenicity

The human adenoviruses are almost exclusively pathogenic for man. Therefore most of the information about adenovirus pathogenesis is derived from experiments with human volunteers. As early as 1947, it was shown that a disease known as *acute respiratory disease* (ARD) could be transferred to volunteers by filtered secretions from sick recruits (Commision on Acute Respiratory Diseases, 1947). Later, GINSBERG et al. (1955) demonstrated that the virus transmitted in this study was an adenovirus closely related to human ad 4. Serotypes ad 1, ad 2, ad 3 and ad 4 were found to induce subclinical infections with antibody response in volunteers, but no respiratory disease was caused by these serotypes when virus grown in HeLa cells was used for inoculation (BELL et al., 1956). On the other hand, ad 1 which had been propagated in embryonic human kidney cells gave rise to pharyngitis (RODEN et al., 1956). Other adenovirus types have been found to induce conjunctivitis and pharyngitis when inoculated in the conjunctival sac (BELL et al., 1956; KASEL et al., 1963 a and b). Thus, it appears that Koch's postulates have been fulfilled for several of the adenovirus types with regard to respiratory infections and conjunctivitis in adults.

Epidemiological studies (for a review see GINSBERG and DINGLE, 1965) have shown that adenoviruses in subgroups B and C differ with regard to diseases they cause in humans. Subgroup C viruses like ad 1, ad 2, ad 5 and ad 6 give rise to respiratory infections in children but cause only rarely infections in adults. These serotypes are also often associated with latent infection in human adenoids and tonsils, and in fact the adenoviruses were originally discovered as passenger virus in cell cultures of adenoids (ROWE et al., 1953 and 1955). Subgroup B viruses, in particular ad 3, ad 7, ad 14 and ad 21, appear to cause outbreaks of fever and pharyngitis, the ARD syndrome, in military recruits and in boarding schools. It is interesting that college students of similar age living in dormitories do not have such outbreaks (KATZ et al., 1959). Ad 4 which is an aberrant member of subgroup C resembles in this respect to subgroup B adenoviruses since it can cause outbreaks of ARD in isolated populations. Epidemic keratoconjunctivitis is another syndrome which is caused by adenoviruses. Ad 8, a member of the subgroup D viruses, has commonly been isolated from patients with epidemic keratoconjunctivitis (JAWETZ, 1957) and this serotype seems also to be responsible for outbreaks of swimming pool conjunctivitis (FUKUMI et al., 1958).

The prevalence of complement-fixing or neutralizing antibodies differs widely in civilian populations (GINSBERG and DINGLE, 1965). Children have a high frequency of antibodies against the subgroup C adenoviruses but only rarely against subgroup B adenoviruses. Young adults usually lack antibodies against ad 4 and ad 7 and this may explain why these two serotypes commonly give rise to epidemics of ARD among military recruits. In the civilian population adenoviruses only rarely cause infections. BRANDT et al. (1969) estimated from a survey of 18,000 children that only 7 per cent of all cases of respiratory disease were associated with adenovirus infection. The most commonly isolated serotypes were ad 1, ad 2, ad 5, ad 6, and less prevalent ad 3 and ad 7. In institutionalized populations adenovirus infections are of importance since they give rise to epidemics. HUEBNER (1959) has correlated some syndromes with the following adenovirus serotypes: Acute respiratory disease (ARD) seems frequently to be connected with ad 4 and ad 7, pharyngoconjunctival fever with ad 3 and ad 7, acute febrile conjunctivitis with ad 3 and ad 7 and epidemic keratoconjunctivitis with ad 8. Pneumonia in adults is sometimes associated with ad 4 and ad 7.

In general, laboratory animals do not seem to be susceptible to infection with the human adenoviruses, but PEREIRA and KELLEY (1957 b) produced a latent ad 5 infection in rabbits. Subgroup C adenoviruses, like ad 1, ad 2, ad 5 and ad 6, have been used to produce pneumonia in young piglets deprived of colostrum (BETTS et al., 1962). Experimental infection of calves with ad 1 (BETTINOTTI and STRAUT, 1966) and subclinical infections of dogs (SINHA et al., 1960) with human adenoviruses have also been reported.

2. Epidemiology and Control of Adenovirus Infection

Adenoviruses are usually recovered from the respiratory tract but are also present in stools and urine (COUCH et al., 1966; NUMAZAKI et al., 1968). In some cases virus has been recovered only from the intestinal tract, although the infection still seems to be spread by respiratory secretions (COUCH et al., 1966).

Shortly after adenoviruses were isolated and the epidemiology established, formalinized vaccines were produced and shown to efficiently control epidemics among military recruits (for a review see HILLEMAN, 1966). The vaccines were later withdrawn because the adenoviruses were cultivated in monkey cells which were contaminated with oncogenic SV 40 virus and later it was also shown that the adenoviruses themselves were oncogenic. Vaccination with attenuated strains of the most prevalent serotypes has been tried and found effective (SELIVANOV et al., 1964). Successful vaccination has also been achieved by oral administration of a live virus enclosed in capsules (CHANOCK et al., 1966; GUTEKUNST et al., 1967). It has long been suggested that vaccines which contain only adenovirus structural proteins should preferably be used for induction of neutralizing antibodies in humans (KASEL et al., 1964). A recent report (COUCH et al., 1973) suggests that both crystalline hexon and crystalline fiber preparations induce high levels of neutralizing antibodies. This vaccine certainly deserves further investigation.

B. Infections in Animals

Several simian serotypes have been found to cause respiratory as well as enteric infection in monkeys and chimpanzees (EUGSTER et al., 1969). Serological and epidemiological studies indicate that bovine adenoviruses may play a role in the etiology of respiratory disease among cattle (DARBYSHIRE and PEREIRA, 1964; ALDASI et al., 1965) and conjunctivitis and keratoconjunctivitis has also been associated with the bovine adenoviruses (WILCOX, 1969). Experimental infection of young calves appears to give rise to respiratory and in some instances intestinal symptoms (ALDASI et al., 1965; DARBYSHIRE et al., 1969). The porcine adenoviruses, in particular types 2 and 3, seem to give rise to subclinical infection of both the tonsils and the lower intestinal tract (SHARPE and JESSET, 1967). Pig embryos were also aborted after inoculation of type 1 in utero (SHARPE, 1967). Equine adenoviruses have been found to cause severe pneumonia which among Arabian foals often is fatal (McCHESNEY et al., 1973). The infectious canine hepatitis (ICH) agent, which was identified as an adenovirus by KAPSENBERG (1959), causes subclinical infection in adult dogs. Puppies, on the other hand, show severe symptoms of fever and gastro-intestinal disturbances and the mortality is 10—25 per cent. In foxes the same agent causes an acute encephalitis with convulsions followed by coma and death within 24 hours. Both formalinized, killed and live attenuated vaccines have been used to control the disease (CABASSO et al., 1958; PIERCY and SELLERS, 1960). The murine adenoviruses may cause a fatal disease when inoculated in suckling mice (HARTLEY and ROWE, 1960), but other strains apparently only cause subclinical infection (MISSAL, 1969). Among the avian adenoviruses Gallus adeno-like (GAL) virus can induce necrosis of the liver in chickens (SHARPLESS and JUNGHERR, 1961) but has not been isolated from animals with a natural disease. Chicken embryo lethal orphan (CELO) virus is the causative agent of quail bronchitis and this condition can be reproduced experimentally (DUBOSE and GRUMBLES, 1959). When inoculated in chick embryos the virus is lethal.

The diseases which so far have been ascribed to human adenoviruses are not severe enough to call for an extensive vaccination program. The fact that several of the human adenoviruses are oncogenic has hampered the development of safe vaccines. In man, prophylaxis should probably be limited to vaccines which are composed of the protein moiety which induces neutralizing antibodies. Promising results have already been obtained with crystalline vaccines (COUCH et al., 1973) and similar vaccines might be used in recruit camps to counteract the regular epidemics of acute respiratory disease at the commencement of the military service. It appears to be less urgent to vaccinate the civilian population since only few cases of respiratory illnesses are caused by adenoviruses. Dangerous adenovirus infections in dogs and calves could be controlled efficiently with live or inactivated vaccines.

XV. Aspects on Adenoviruses as a Tool in Cell Biology

RNA synthesis during adenovirus reproduction shows many similarities with the synthesis of host RNA in uninfected cells. Adenoviruses probably do not carry their own enzymes for synthesis of RNA or DNA, and no virus-specific polymerases have been detected during infection. This would imply that the virus utilizes host enzymes for transcription and replication of its genome. The efficiency of virus reproduction indicates, however, that adenoviruses have acquired functions, by which it can direct the host cell enzymes to preferentially replicate and transcribe viral DNA and to take over the protein synthesizing machinery of the host cell. An understanding of regulatory mechanisms for macromolecular synthesis during adenovirus infection might inform us about control mechanisms in eukaryotic cells in general.

The ultimate result of early gene expression is the triggering of viral DNA replication, which in turn is necessary to switch on the expression of late viral genes. The early protein(s) which are necessary to initiate viral DNA synthesis has not been identified. A protein synthesized towards the end of the early phase may govern this shift and the virus-specific DNA binding proteins described by van der Vliet and Levine (1973) are possible candidates. Several important questions remain to be answered concerning the mechanisms controlling adenovirus transcription at different times after infection. It is well known that, if viral DNA replication is prevented, the late mRNA is not made. This could suggest that the appropriate DNA template for synthesis of late mRNA is provided through replication of the virus DNA. Alternatively, the onset of viral DNA replication leads to changes in the specificity of mRNA polymerase or of enzymes which are responsible for processing of RNA.

Studies on adenovirus DNA replication might also be rewarding. Adenoviruses differ from most other microorganisms in the structure of the replicating DNA and in the lack of requirement for continuous protein synthesis after DNA replication has begun (Horwitz et al., 1973). The isolation of adenovirus mutants, which are temperature-sensitive for viral DNA replication, and the development of a nuclear system for *in vitro* DNA synthesis in which DNA synthesis can be studied under different conditions, are two important advances which soon could lead to the identification of the factor(s) required for replication of adenovirus DNA.

Another interesting aspect of adenovirus reproduction is the suppression of the host protein synthesis late in infection. Some regulatory mechanism has to operate at the translational level causing the replacement of the host mRNA by viral mRNA in protein synthesizing polyribosomes. Since the host mRNA is long-lived and no enhanced degradation of host mRNA has been observed after infection, this effect may be caused by specific initiation of viral messenger RNA. Such effects could be due to the structure of the viral mRNA itself or specific viral initiation factors. A possible candidate for such a function is the polypeptide which is found associated with messenger RNA exclusively in adenovirus infected cells (Lindberg and Sundquist, 1974). Adenovirus infection in monkey cells provides an interesting possibility to study translational regulation. Infection of monkey cells with human adenoviruses leads to synthesis of viral DNA and both early and late viral mRNA, but late viral mRNA cannot be efficiently translated unless an unidentified SV 40 function is concurrently expressed. This helper function can be provided by SV 40 virus, adenovirus-SV 40 hybrids and monkey cells which are transformed by SV 40. Studies on non-defective adenovirus-SV 40 hybrids suggest that the early SV 40 U-antigen may be related to this regulatory protein which facilitates adenovirus translation. A specific polypeptide band present in cells infected with SV 40 or the hybrid virus ad 2^+ND_1 has already been identified (López-Revilla and Walter, 1973). Further studies on this system should provide valuable information on regulatory mechanisms for adenovirus translation.

Many adenovirus serotypes are oncogenic and adenoviruses in general seem to be able to transform cells *in vitro*. Adenovirus transformed cells have many properties in common with tumor cells and also cells transformed by other viruses. Since cell transformation by viruses involves addition of a small amount of extra genetic information, it may ultimately be possible to correlate certain viral gene products with changes in growth properties of cells *in vitro*. It is of great interest that adenovirus transformation can be accomplished with naked DNA and that only a minor part of the genome seems to be required both for transformation and for maintenance of the transformed state (Graham and Van der Eb, 1973b; Sharp et al., 1974b; Graham et al., 1974; Gallimore et al., 1974). The segment of the ad 2 chromosome which is necessary for cell transformation has already been identified (Graham et al., 1974; Gallimore et al., 1974). This DNA segment could be used for selection of mRNA which could be translated in an *in vitro* protein synthesizing system. Thus, it might be possible in the near future to identify the protein(s) which is required for transformation.

The covalent insertion of viral gene(s) into cellular DNA in transformed cells constitutes a convenient tool to study synthesis and processing of RNA in eukaryotic cells, since the viral RNA can easily be recognized by hybridization to viral DNA. Important observations have already been made with the aid of transformed cells. Thus, it was demonstrated in cells transformed by SV 40 and adenoviruses that viral RNA sequences are present in larger molecules in the nucleus than in the polyribosomes (Lindberg and Darnell, 1970; Tonegawa et al., 1970; Green et al., 1970). Subsequently, it has been shown that the nuclear RNA in transformed cells contains covalently linked host cell and viral RNA sequences (Tseui et al., 1972; Wall et al., 1973). In contrast, mRNA in the cytoplasm contains only viral sequences which appear to be derived from the 3' end of giant nuclear RNA molecules (Bachenheimer, 1974).

Important questions, which still remain to be answered, are whether transcription in eukaryotic cells is controlled by specific initiation and termination or by post-transcriptional cleavage and degradation.

In conclusion, the adenoviruses may turn out to be more important as tools to unravel the mechanisms for macromolecular synthesis in mammalian cells than as agents causing human diseases. Thus, these viruses born concurrently with molecular virology in 1953 have and will probably continue to serve as important tools for further progress in this field.

Acknowledgements

The authors are indepted to several colleagues for supplying preprints and unpublished information for this review. Our collaborators Drs. B. Edvardsson, E. Everitt, K. Johansson, T. Linné, T. Persson, L. Prage, C. Tibbetts, J. Saborio, B. Sundquist, and B. Vennström have contributed both ideas and experiments to this review. Drs. C. Anderson, A. Bartha, S. Baum, J. Darnell, W. Doerfler, C. F. Garon, T. Kelly, A. Lewis, H. Raskas, R. Roberts, J. Rose, W. Russell, J. Sambrook, P. Sharp, J. Sussenbach, H. Westphal, E. Winnacker, and J. Williams have kindly provided information on unpublished work and work about to be published. Dr. J. Hazelbauer provided valuable editorial advice.

Most of our gratitude goes, however, to our excellent secretary Mrs. Berit Nordensved, who without despair typed three versions of this book. Additional secretarial aid by Mrs. Gunvor Lindman is also appreciated.

Experiments carried out in the laboratory of the authors were supported by grants from the Swedish Cancer Society, the Swedish Medical and Natural Science Research Councils, and the Wallenberg Foundation.

References[1]

ÁLDÁSI, P., L. CSONTOS, and A. BARTHA: Pneumo-Enteritis in calves caused by adenoviruses. Acta vet. Acad. Sci. hung. **15**, 167—175 (1965).

ALTSTEIN, A. D., and N. N. DODONOVA: Interaction between human and simian adenoviruses in simian cells: Complementation, phenotypic mixing and formation of monkey cell "adapted" virions. Virology **35**, 248—254 (1968).

ALTSTEIN, A. D., O. F. SÁRYCHEVA, and N. N. DODONOVA: Transforming activity of green monkey SA 7 (C 8) adenovirus in tissue culture. Science **158**, 1455—1456 (1967).

ANDERSON, C. W., P. R. BAUM, and R. F. GESTELAND: Processing of adenovirus 2-induced proteins. J. Virol. **12**, 241—252 (1973).

ANDERSON, C. W., J. B. LEWIS, J. F. ATKINS, and R. F. GESTELAND: Cell-free synthesis of adenovirus 2-specific proteins programmed by fractionated mRNA: A comparison of polypeptide product size and messenger RNA lengths. Proc. nat. Acad. Sci. (Wash.) **71**, 2756—2760 (1974).

ANDREWES, C., and H. G. PEREIRAS: Adenoviruses. In: Viruses of Vertebrates, 3rd ed., pp. 309—328. Baillière & Tindall, 1972.

AVIV, H., and P. LEDER: Purification of biologically active globin messenger RNA by chromatography on oligo thymidylic acid-cellulose. Proc. nat. Acad. Sci. (Wash.) **69**, 1408—1412 (1972).

BABLANIAN, R., and W. C. RUSSELL: Adenovirus polypeptide synthesis in the presence of non-replicating poliovirus. J. gen. Virol. **24**, 261—279 (1974).

BACHENHEIMER, personal communication (1974).

BAUER, H., und R. WIGAND: Eigenschaften der Adenovirus-Hämagglutinine. Z. Hyg. Infekt.-Kr. **149**, 96—113 (1963).

BAUM, S. G., and R. I. FOX: Human adenovirus infection in monkey cells. An example of host restriction at a step late in replication. Cold Spr. Harb. Symp. quant Biol. **39**, 567—573 (1974).

BAUM, S. G., W. H. WIESE, and P. R. REICH: Studies on the mechanism of enhancement of adenovirus 7 infection in African green monkey cells by simian virus 40: Formation of adenovirus-specific RNA. Virology **34**, 373—376 (1968).

BAUM, S., M. HORWITZ, and J. MAIZEL, JR.: Studies on the mechanism of enhancement of human adenovirus infection in monkey cells by simian virus 40. J. Virol. **10**, 211—219 (1972).

---

[1] The survey of the literature pertaining to this review was completed in June, 1974.

BAUM, S. G., P. R. REICH, R. J. HUEBNER, W. P. ROWE, and S. M. WEISSMAN: Biophysical evidence for linkage of adenovirus and SV 40 DNAs in adenovirus 7-SV 40 hybrid particles. Proc. nat. Acad. Sci. (Wash.) **56**, 1509—1515 (1966).

BEARDMORE, W. B., M. J. HAVLICK, A. SERAFINI, and I. W. MCLEAN, JR.: Interrelationship of adenovirus (type 4) and papovavirus (SV-40) in monkey kidney cell cultures. J. Immunol. **95**, 422—435 (1965).

BÉLÁDI, I.: Adenoviruses. In: Strains of Human Viruses. (M. MAJER and S. A. PLOTKIN, eds.), pp. 1—19. Basel: S. Karger, 1972.

BELL, J. A., T. G. WARD, R. J. HUEBNER, W. P. ROWE, G. SUSKIND, and R. S. PAFFENBARGER: Studies of adenoviruses (APC) in volunteers. Amer. J. publ. Hlth **46**, 1130—1146 (1956).

BELLETT, A. J. D., and H. B. YOUNGHUSBAND: Replication of the DNA of chick embryo lethal orphan virus. J. molec. Biol. **72**, 691—709 (1972).

BELLO, L. J., and H. S. GINSBERG: Inhibition of host protein synthesis in type 5 adenovirus-infected cells. J. Virol. **1**, 843—850 (1967).

BELLO, L. J., and H. S. GINSBERG: Relationship between deoxyribonucleic acid-like ribonucleic acid synthesis and inhibition of host protein synthesis in type 5 adenovirus-infected KB cells. J. Virol. **3**, 106—113 (1969).

BERMAN, L. D.: On the nature of transplantation immunity in the adenovirus tumor system. J. exp. Med. **125**, 983—1000 (1967).

BERMAN, L. D., and W. P. ROWE: A study of the antigens involved in adenovirus 12 tumorigenesis by immunodiffusion techniques. J. exp. Med. **121**, 955—967 (1965).

BERNHARD, W., N. GRANBOULAN, G. BARSKI et P. TURNER: Essais de cytochimie ultrastructurale. Digestion de virus sur coupes ultrafines. C. R. Acad. Sci. (Paris) **252**, 202—204 (1961).

BETTINOTTI, C. M., und O. C. STRAUB: Experimentelle Infektionen von Rindern mit humanem Adenovirus (Typ 1). Zbl. Bakt. I. Abt. Orig. **199**, 427—431 (1966).

BETTS, A. O., A. R. JENNINGS, P. H. LAMONT, and Z. PAGE: Inoculation of pigs with adenoviruses of man. Nature (Lond.) **193**, 45—46 (1962).

BHADURI, S., H. RASKAS, and M. GREEN: A procedure for the preparation of milligram quantities of adenovirus messenger RNA. J. Virol. **10**, 1126—1129 (1972).

BIBRACK, B.: Untersuchungen über serologische Einordnung von 9 in Bayern aus Schweinen isolierten Adenovirus-Stämmen. Zbl. Vet.-Med. B. **16**, 327—334 (1969).

BIRNSTIEL, M. L., E. S. WEINBERG, and M. L. PARDUE: Evolution of 9S mRNA sequences. In: Molecular Cytogenetics. (B. A. HAMKALO and J. PAPACONSTANTINOU, eds.), pp. 75—93. New York: Plenum Press, 1973.

BISERTE, G., J. SAMAILLE, M. DAUTREVAUX, P. BOULANGER, P. SAUTIÈRE, J. RINGEL, et R. WAROCQUIER: Composition en acides aminés de l'antigène de structure A de l'adénovirus 5. C. R. Acad. Sci. (Paris) **263**, 1648—1649 (1966).

BLACKLOW, N. R., M. D. HOGGAN, J. B. AUSTIN, and W. P. ROWE: Observations on two new adenovirus serotypes with unusual antigenic characteristics. Amer. J. Epidem. **90**, 501—505 (1969).

BLOBEL, G.: A protein of molecular weight 78,000 bound to the polyadenylate region of eukaryotic messenger RNAs. Proc. nat. Acad. Sci. (Wash.) **70**, 924—928 (1973).

BOULANGER, P. A., and F. PUVION: Large-scale preparation of soluble adenovirus hexon, penton and fiber antigens in highly purified form. Europ. J. Biochem. **39**, 37—42 (1973).

BOULANGER, P. A., P. FLAMENCOURT, and G. BISERTE: Isolation and comparative chemical study of structure proteins of the adenovirus 2 and 5: Hexon and fiber antigens. Europ. J. Biochem. **10**, 116—131 (1969).

BOULANGER, P., G. TORPIER, and G. BISERTE: Investigations on intranuclear paracrystalline inclusions induced by adenovirus 5 in KB cells. J. gen. Virol. **6**, 329—332 (1970).

BOULANGER, P. A., G. TORPIER, and A. RIMSKY: Crystallographic study of intranuclear adenovirus type 5 crystals. Intervirology **2**, 56—62 (1973).

BRANDT, C. D., H. W. KIM, A. J. VARGOSKO, B. C. JEFFRIES, J. O. ARROBIO, B. RINDGE, R. H. PARROTT, and R. M. CHANOCK: Infections in 18,000 infants and children in a

controlled study of respiratory tract disease. I. Adenovirus pathogenicity in relation to serologic type and illness syndrome. Amer. J. Epidem. **90**, 484—500 (1969).
BRESNICK, E., and F. RAPP: Thymidine kinase activity in cells abortively and productively infected with human adenoviruses. Virology **34**, 799—802 (1968).
BRONITKI, A., R. DEMETRESCU, G. POPESCU, and A. MALIAN: Isolation of adenovirus from a human case of pulmonary carcinoma. Acta virol. **8**, 472 (1964).
BROWN, D. T., and B. T. BURLINGHAM: Penetration of host cell membranes by adenovirus 2. J. Virol. **12**, 386—396 (1973).
BRUTLAG, D., R. SCHEKMAN, and A. KORNBERG: A possible role for RNA polymerase in the initiation of M13 DNA synthesis. Proc. nat. Acad. Sci. (Wash.) **68**, 2826 to 2829 (1971).
BURGER, H., and W. DOERFLER: Intracellular forms of adenovirus deoxyribonucleic acid III. Integration of the deoxyribonucleic acid of adenovirus type 2 into host deoxyribonucleic acid in productively infected cells. J. Virol. **13**, 975—992 (1974).
BURLINGHAM, B., and W. DOERFLER: Three size classes of adenovirus intranuclear deoxyribonucleic acid. J. Virol. **7**, 707—719 (1971).
BURLINGHAM, B., and W. DOERFLER: An endonuclease in cells infected with adenovirus and associated with adenovirions. Virology **48**, 1—13 (1972).
BURLINGHAM, B., W. DOERFLER, U. PETTERSSON, and L. PHILIPSON: Adenovirus endonuclease: Association with the penton of adenovirus type 2. J. molec. Biol. **60**, 45—64 (1971).
BURNETT, J. P., and J. A. HARRINGTON: Simian adenovirus SA7 DNA: Chemical, physical and biological studies. Proc. nat. Acad. Sci. (Wash.) **60**, 1023—1029 (1968a).
BURNETT, J. P., and J. A. HARRINGTON: Infectivity associated with simian adenovirus type SA7 DNA. Nature (Lond.) **220**, 1245—1246 (1968 b).
BURNS, W. H., and P. H. BLACK: Induction experiments with adenovirus and polyoma virus transformed cell lines. Int. J. Cancer **4**, 204—211 (1969 a).
BURNS, W. H., and P. H. BLACK: Analysis of SV40-induced transformation of hamster kidney tissue *in vitro*. VI. Characteristics of mitomycin C induction. Virology **39**, 625—634 (1969 b).
BUTEL, J. S., and F. RAPP: Complementation between a defective monkey cell adapting component and human adenovirus in simian cells. Virology **31**, 573—584 (1967).
CABASSO, V. J., M. R. STEBBINS, and J. M. AVAMPATO: A bivalent live virus vaccine against canine distemper (CD) and infectious canine hepatitis (ICH). Proc. Soc. exp. Biol. (N.Y.) **99**, 46—51 (1958).
CAFFIER, H., and M. GREEN: Adenovirus proteins. III. Cell-free synthesis of adenovirus proteins in cytoplasmic extracts of KB cells. Virology **46**, 98—105 (1971).
CAFFIER, H., H. J. RASKAS, J. T. PARSONS, and M. GREEN: *In vitro* and *in vivo* initiation of mammalian viral protein synthesis by methionyl-tRNA. Nature (New Biol.) **229**, 239—241 (1971).
CASTO, B. C.: Effects of ultraviolet irradiation on the transforming and plaque-forming capacities of simian adenovirus SA7. J. Virol. **2**, 641—642 (1968 a).
CASTO, B. C.: Adenovirus transformation of hamster embryo cells. Effect of animal age and sex on susceptibility to transformation. Bact. Proc. **68**, 176 (1968 b).
CASTO, B. C.: Transformation of hamster embryo cells and tumor induction in newborn hamsters by simian adenovirus SV11. J. Virol. **3**, 513—519 (1969).
CASTO, B. C.: Biologic parameters of adenovirus transformation. Progr. exp. Tumor Res. (Basel) **18**, 166—198 (1973).
CHAMBON, P., F. GISSINGER, J. L. MANDEL, JR., C. KEDINGER, M. GNIAZDOWSKI, and M. MEIHLAC: Purification and properties of calf thymus DNA-dependent RNA polymerase A and B. Cold Spr. Harb. Symp. quant. Biol. **35**, 693—707 (1970).
CHANOCK, R. M., W. LUDWIG, R. J. HUEBNER, T. R. CATE, and L.-W. CHU: Immunization by selective infection with type 4 adenovirus grown in human diploid tissue cultures. J. Amer. med. Ass. **195**, 445—453 (1966).
CHARDONNET, Y., and S. DALES: Early events in the interaction of adenovirus with HeLa cells. I. Penetration of type 5 and intracellular release of the DNA genome. Virology **40**, 462—477 (1970 a).

CHARDONNET, Y., and S. DALES: Early events in the interaction of adenovirus with HeLa cells. II. Comparative observation on the penetration of type 1, 5, 7, 12. Virology **40**, 478—485 (1970 b).

CHARDONNET, Y., and S. DALES: Early events in the interaction of adenoviruses with HeLa cells. III. Relationship between an ATPase activity in nuclear envelopes and transfer of core material: a hypothesis. Virology **48**, 342—359 (1972).

CHARDONNET, Y., L. GAZZOLO, and B. POGO: Effect of α-aminitin on adenovirus 5 multiplication. Virology **48**, 300—304 (1972).

CLARKE, M. C., H. B. A. SHARPE, and J. B. DERBYSHIRE: Some characteristics of three porcine adenoviruses. Arch. ges. Virusforsch. **21**, 91—97 (1967).

Commision on Acute Respiratory Diseases: Experimental transmission of minor respiratory illness to human volunteers by filter-passing agents. I. Demonstration of two types of illness characterized by long and short incubation periods and different clinical features. J. clin. Invest. **26**, 957—973 (1947).

CORNICK, G., P. B. SIGLER, and H. S. GINSBERG: Characterization of crystals of adenovirus type 5 hexon. J. molec. Biol. **57**, 397—401 (1971).

CORNICK, G., P. B. SIGLER, and H. S. GINSBERG: Mass of protein in the asymmetric unit of hexon crystals — a new method. J. molec. Biol. **73**, 533—537 (1973).

COUCH, R. B., T. R. CATE, W. F. FLEET, P. J. GERONE, and V. KNIGHT: Aerosol-induced adenoviral illness resembling the naturally occurring illness in military recruits. Ann. Rev. resp. Dis. **93**, 529—535 (1966).

COUCH, R. B., J. A. KASEL, H. G. PEREIRA, A. T. HAASE, and V. KNIGHT: Induction of immunity in man by crystalline adenovirus type 5 capsid antigens. Proc. Soc. exp. Biol. (N.Y.) **143**, 905—910 (1973).

CRAIG, E. A., J. TAL, T. NISHIMOTO, M. MCGROGAN, S. ZIMMER, and H. J. RASKAS: RNA transcription in cultures productively infected with adenovirus 2. Cold Spr. Harb. Symp. quant. Biol. **39**, 483—493 (1974).

CROWTHER, R. A., and R. M. FRANKLIN: The structure of the groups of nine hexons from adenovirus. J. molec. Biol. **68**, 181—184 (1972).

CRUMPACKER, C. S., M. J. LEVIN, W. H. WIESE, A. M. LEWIS, JR., and W. P. ROWE: The adenovirus type 2-simian virus 40 hybrid population: Evidence for a hybrid deoxyribonucleic acid molecule and the absence of adenovirus-encapsidated circular simian virus 40 deoxyribonucleic acid. J. Virol. **6**, 788—794 (1970).

DALES, S.: An electron microscope study of the early association between two mammalian viruses and their hosts. J. Cell Biol. **13**, 303—322 (1962).

DALES, S., and Y. CHARDONNET: Early events in the interaction of adenoviruses with HeLa cells. IV. Association with microtubules and the nuclear pore complex during vectorial movement of the inoculum. Virology **56**, 465—483 (1973).

DARBYSHIRE, J. H.: Oncogenicity of bovine adenovirus type 3 in hamsters. Nature (Lond.) **211**, 102 (1966).

DARBYSHIRE, J. H.: Bovine adenoviruses. Progr. exp. Tumor Res. (Basel) **18**, 56—66 (1973).

DARBYSHIRE, J. H., and H. G. PEREIRA: An adenovirus precipitating antibody present in some sera of different animal species and its association with bovine respiratory disease. Nature (Lond.) **201**, 895—897 (1964).

DARBYSHIRE, J. H., D. A. KINCH, and A. R. JENNINGS: Experimental infection of calves with bovine adenovirus types 1 and 2. Res. vet. Sci. **10**, 39—45 (1969).

DARNELL, J. E., R. WALL, and R. TUSHINSKI: An adenylic acid-rich sequence in messenger RNA of HeLa cells and its possible relationship to reiterated sites in DNA. Proc. nat. Acad. Sci. (Wash.) **68**, 1321—1325 (1971 a).

DARNELL, J. E., W. R. JELINEK, and G. R. MOLLOY: Biogenesis of mRNA: Genetic regulation in mammalian cells. Science **181**, 1215—1221 (1973).

DARNELL, J. E., L. PHILIPSON, R. WALL, and M. ADESNIK: Polyadenylic acid sequences: Role in conversion of nuclear RNA into messenger RNA. Science **174**, 507—510 (1971 b).

DARNELL, J. E., R. WALL, M. ADESNIK, and L. PHILIPSON: The formation of messenger RNA in HeLa cells by post-transcriptional modification of nuclear RNA. In:

Molecular Genetics and Developmental Biology (M. SUSSMAN, ed.), pp. 201—225. Prentice-Hall Inc., 1972.

DIAMANDOPOULUS, G. TH.: Comparison of the *in vitro* cytology of hamster embryo cell lines transformed "spontaneously" or by SV 40. Amer. J. Path. **52**, 633—647 (1968 a).

DIAMANDOPOULUS, G. TH.: Histopathology of sarcomas induced in hamsters by clones of *in vitro* SV 40-transformed homologous heart cells. Amer. J. Path. **53**, 753—767 (1968 b).

DOERFLER, W.: The fate of the DNA of adenovirus type 12 in baby hamster kidney cells. Proc. nat. Acad. Sci. (Wash.) **60**, 636—646 (1968).

DOERFLER, W.: Non-productive infection of baby hamster kidney cells (BHK 21) with adenovirus type 12. Virology **38**, 587—606 (1969).

DOERFLER, W.: Integration of the DNA of adenovirus type 12 into the DNA of baby hamster kidney cells. J. Virol. **6**, 652—666 (1970).

DOERFLER, W., and A. K. KLEINSCHMIDT: Denaturation pattern of the DNA of adenovirus type 2 as determined by electron microscopy. J. molec. Biol. **50**, 579—593 (1970).

DOERFLER, W., and U. LUNDHOLM: Absence of replication of the DNA of adenovirus type 12 in BHK 21 cells. Virology **40**, 754—756 (1970).

DOERFLER, W., W. HELLMAN, and A. K. KLEINSCHMIDT: The DNA of adenovirus type 12 and its denaturation pattern. Virology **47**, 507—512 (1972).

DOERFLER, W., U. LUNDHOLM, U. RENSING, and L. PHILIPSON: Intracellular forms of adenovirus deoxyribonucleic acid. II. Isolation in dye-buoyant density gradients of a deoxyribonucleic acid-ribonucleic acid complex from KB cells infected with adenovirus type 2. J. Virol. **12**, 793—807 (1973).

DUBOSE, R. T., and L. C. GRUMBLES: The relationship between quail bronchitis virus and chicken embryo lethal orphan virus. Avian Dis. **3**, 231—343 (1959).

DULBECCO, R., and M. VOGT: Some problems of animal virology as studied by the plaque technique. Cold Spr. Harb. Symp. quant. Biol. **18**, 273—279 (1953).

DUNN, A. R., P. H. GALLIMORE, K. W. JONES, and J. K. MCDOUGALL: In situ hybridization of adenovirus RNA and DNA. II. Detection of adenovirus specific DNA in transformed and tumor cells. Int. J. Cancer **11**, 628—636 (1973).

EASTON, J. M., and C. W. HIATT: Possible incorporation of SV 40 genome within capsid proteins of adenovirus 4. Proc. nat. Acad. Sci. (Wash.) **54**, 1100—1104 (1965).

EDMONDS, M., and M. G. CARAMELA: The isolation and characterization of adenosine-monophosphate rich polynucleotides synthesized by Ehrlich ascites cells. J. biol. Chem. **244**, 1314—1324 (1969).

EDMONDS, M., M. H. VAUGHAN, and N. NAKAZOTO: Polyadenylic acid sequences in the heterogeneous nuclear RNA and rapidly labeled polyribosomal RNA of HeLa cells: Possible evidence for a precursor relationship. Proc. nat. Acad. Sci. (Wash.) **68**, 1336—1340 (1971).

ELICEIRI, G. L.: Ribosomal RNA synthesis after infection with adenovirus type 2. Virology **56**, 604—607 (1973).

ELLENS, D. J., J. S. SUSSENBACH, and H. S. JANSZ: Studies on the mechanism of replication of adenovirus DNA. III. Electron microscopy of replicating DNA. Virology **61**, 427—442 (1974).

ENDERS, J. F., J. A. BELL, J. H. DINGLE, T. FRANCIS, JR., M. R. HILLEMAN, R. J. HUEBNER, and A. M.-M. PAYNE: Adenoviruses: Group name proposed for new respiratory tract viruses. Science **124**, 119—120 (1956).

ENSINGER, M. J., and H. S. GINSBERG: Selection and preliminary characterization of temperature-sensitive mutants of type 5 adenovirus. J. Virol. **10**, 328—339 (1972).

EPSTEIN, M. A.: Observations on the fine structure of type 5 adenovirus. J. biophys. biochem. Cytol. **6**, 523—526 (1959).

EPSTEIN, M. A., S. J. HOLT, and A. K. POWELL: The fine structure and composition of type 5 adenovirus: An integrated electron microscopial and cytochemical study. Brit. J. exp. Path. **41**, 567—576 (1960).

ERON, L., and H. WESTPHAL: Cell-free translation of highly purified adenovirus messenger RNA. Proc. nat. Acad. Sci. (Wash.) **71**, 3385—3389 (1974).

ERON, L., R. CALLAHAN, and H. WESTPHAL: Cell-free synthesis of adenovirus coat proteins. J. biol. Chem. **249**, 6331—6338 (1974 a).

ERON, L., H. WESTPHAL, and R. CALLAHAN: *In vitro* synthesis of adenovirus core proteins. J. Virol. **14**, 375—383 (1974 b).

EUGSTER, A. K., S. S. KALTER, C. S. KIM, and M. E. PINKERTON: Isolation of adenoviruses from baboons (*Papio* sp.) with respiratory and enteric infections. Arch. ges. Virusforsch. **26**, 260—270 (1969).

EVERETT, S. F., and H. S. GINSBERG: A toxin-like material separable from type 5 adenovirus particles. Virology **6**, 770—771 (1958).

EVERITT, E., and L. PHILIPSON: Structural proteins of adenoviruses. XI. Purification of three low molecular weight proteins of adenovirus type 2 and their synthesis during productive infection. Virology **62**, 253—269 (1974).

EVERITT, E., B. SUNDQUIST, and L. PHILIPSON: Mechanism of arginine requirement for adenovirus synthesis. I. Synthesis of structural proteins. J. Virol. **8**, 742—753 (1971).

EVERITT, E., B. SUNDQUIST, U. PETTERSSON, and L. PHILIPSON: Structural proteins of adenoviruses. X. Isolation and topography of low molecular weight antigens from the virion of adenovirus type 2. Virology **52**, 130—147 (1973).

FELDMAN, L. A., and F. RAPP: Inhibition of adenovirus replication by 1-ß-D-arabinofuranosylcytosine. Proc. Soc. exp. Biol. (N.Y.) **122**, 243—247 (1966).

FELDMAN, L. A., J. S. BUTEL, and F. RAPP: Interaction of a simian papovavirus and adenovirus. I. Induction of adenovirus tumor antigen during abortive infection of simian cells. J. Bact. **91**, 813—818 (1966).

FLANAGAN, J. F., and H. S. GINSBERG: Synthesis of virus-specific polymers in adenovirus-infected cells: Effect of 5-fluorodeoxyuridine. J. exp. Med. **116**, 141—157 (1962).

FOGEL, M., and L. SACHS: The activation of virus synthesis in polyoma-transformed cells. Virology **37**, 327—334 (1969).

FOGEL, M., and L. SACHS: Induction of virus synthesis in polyoma-transformed cells by ultraviolet light and mitomycin C. Virology **40**, 174—177 (1970).

FOX, R., and S. BAUM: Synthesis of viral ribonucleic acid during restricted adenovirus infection. J. Virol. **10**, 220—227 (1972).

FOX, R., and S. BAUM: Posttranscriptional block to adenovirus replication in nonpermissive monkey cells. Virology **60**, 45—53 (1974).

FRANKLIN, R. M., U. PETTERSSON, K. ÅKERVALL, B. STRANDBERG, and L. PHILIPSON: Structural proteins of adenoviruses. V. On the size and structure of the adenovirus type 2 hexon. J. molec. Biol. **57**, 383—395 (1971 a).

FRANKLIN, R. M., S. C. HARRISON, U. PETTERSSON, C. I. BRÄNDÉN, P. E. WERNER, and L. PHILIPSON: Structural studies on the adenovirus hexon. Cold Spr. Harb. Symp. quant. Biol. **36**, 503—510 (1971 b).

FREEMAN, A. E., P. H. BLACK, R. WOLFORD, and R. J. HUEBNER: Adenovirus type 12-rat embryo transformation system. J. Virol. **1**, 362—367 (1967 a).

FREEMAN, A. E., P. H. BLACK, E. A. VANDERPOOL, P. H. HENRY, J. B. AUSTIN, and R. J. HUEBNER: Transformation of primary rat embryo cells by adenovirus type 2. Proc. nat. Acad. Sci. (Wash.) **58**, 1205—1212 (1967 b).

FREEMAN, A. E., E. A. VANDERPOOL, P. H. BLACK, H. C. TURNER, and R. J. HUEBNER: Transformation of primary rat embryo cells by a weakly oncogenic adenovirus type 3. Nature (Lond.) **216**, 171—173 (1967 c).

FRIEDMAN, D. L., and G. C. MUELLER: A nuclear system for DNA-replication from synchronized HeLa cells. Biochim. biophys. Acta (Amst.) **161**, 455—468 (1968).

FRIEDMAN, M. P., M. J. LYONS, and H. S. GINSBERG: Biochemical consequences of type 2 adenovirus and SV 40 double infection of African green monkey kidney cells. J. Virol. **5**, 586—597 (1970).

FUJINAGA, K., and M. GREEN: The mechanism of viral carcinogenesis by DNA mammalian viruses: Viral-specific RNA in polyribosomes of adenovirus tumor and transformed cells. Proc. nat. Acad. Sci. (Wash.) **55**, 1567—1574 (1966).

FUJINAGA, K., and M. GREEN: Mechanism of viral carcinogenesis by deoxyribonucleic acid mammalian viruses: IV. Related virus-specific ribonucleic acids in tumor cells induced by "highly" oncogenic adenovirus type 12, 18 and 31. J. Virol. **1**, 576—582 (1967 a).

FUJINAGA, K., and M. GREEN: Mechanism of viral carcinogenesis by DNA mammalian viruses: II. Viral-specific RNA in tumor cells induced by "weakly" oncogenic human adenoviruses. Proc. nat. Acad. Sci. (Wash.) **57**, 806—812 (1967 b).

FUJINAGA, K., and M. GREEN: Mechanism of viral carcinogenesis by DNA mammalian viruses: V. Properties of purified viral-specific RNA from human adenovirus-induced tumor cells. J. molec. Biol. **31**, 63—73 (1968).

FUJINAGA, K., and M. GREEN: Mechanism of viral carcinogenesis by DNA mammalian viruses: VII. Viral genes transcribed in adenovirus type 2 infected and transformed cells. Proc. nat. Acad. Sci. (Wash.) **65**, 375—382 (1970).

FUJINAGA, K., M. PIÑA, and M. GREEN: The mechanism of viral carcinogenesis by DNA mammalian viruses: VI. A new class of virus-specific RNA molecules in cells transformed by group C human adenoviruses. Proc. nat. Acad. Sci. (Wash.) **64**, 255—262 (1969).

FUKUMI, H., F. NISHIKAWA, U. KURIMOTO, J. H. INOUE, USUI, and T. HIRAYAMA: Epidemiological studies of an outbreak of epidemic keratoconjunctivitis in Ogaki city and its vicinity, gifu prefecture in 1957. Jap. J. med. Sci. Biol. **11**, 467—481 (1958).

GALLIMORE, P. H.: Tumour production in immunosuppressed rats with cells transformed *in vitro* by adenovirus type 2. J. gen. Virol. **16**, 99—102 (1972).

GALLIMORE, P. H., P. A. SHARP, and J. SAMBROOK: Viral DNA in transformed cells. II. A study of the sequences of adenovirus 2 DNA in nine lines of transformed rat cells using specific fragments of the viral genome. J. molec. Biol. **89**, 49—72 (1974).

GARON, C. F., K. BERRY, and J. ROSE: A unique form of terminal redundancy in adenovirus DNA molecules. Proc. nat. Acad. Sci. (Wash.) **69**, 2391—2395 (1972).

GARON, C. F., K. W. BERRY, J. C. HIERHOLZER, and J. A. ROSE: Mapping of base sequence heterologies between genomes from different adenovirus serotypes. Virology **54**, 414—426 (1973).

GELB, L. D., D. E. KOHNE, and M. A. MARTIN: Quantitation of simian virus 40 sequences in African green monkey, mouse and virus-transformed cell genomes. J. molec. Biol. **57**, 129—145 (1971).

GERBER, P.: Studies on the transfer of subviral infectivity from SV 40-induced hamster tumor cells to indicator cells. Virology **28**, 501—509 (1966).

GILDEN, R. V., J. KERN, Y. K. LEE, F. RAPP, J. L. MELNICK, J. L. RIGGS, E. H. LENNETTE, B. ZBAR, H. J. RAPP, H. C. TURNER, and R. J. HUEBNER: Serologic surveys of human cancer patients for antibody to adenovirus T antigens. Amer. J. Epidem. **91**, 500—509 (1970).

GILEAD, Z., and H. S. GINSBERG: Characterization of a tumorlike antigen in type 12 and type 18 adenovirus-infected cells. J. Bact. **90**, 120—125 (1965).

GILEAD, Z., and H. S. GINSBERG: Characterization of the tumorlike (T) antigen induced by type 12 adenovirus. I. Purification of the antigen from infected KB cells and a hamster cell line. J. Virol. **2**, 7—14 (1968 a).

GILEAD, Z., and H. S. GINSBERG: Characterization of a tumorlike (T) antigen induced by type 12 adenoviruses. II. Physical and chemical properties. J. Virol. **2**, 15—20 (1968 b).

GINSBERG, H. S.: Biochemistry of adenovirus infection. In: The Biochemistry of Viruses (H. B. LEVY, ed.), pp. 329—359. New York: Marcel Dekker, 1969.

GINSBERG, H. S., and J. H. DINGLE: The adenovirus group. In: Viral and Rickettsial Infections of Man (F. HORSFALL and I. TAMM, eds.), 4th ed., pp. 860—891. J. B. Lippincott, 1965.

GINSBERG, H. S., L. J. BELLO, and A. J. LEVINE: Control of biosynthesis of host macromolecules in cells infected with adenovirus. In: The Molecular Biology of Viruses (J. S. COLTER and W. PARANCHYCH, eds.), pp. 547—572. New York: Academic Press, 1967.

GINSBERG, H. S., H. G. PEREIRA, R. C. VALENTINE, and W. C. WILCOX: A proposed terminology for the adenovirus antigens and virion morphological subunits. Virology **28**, 782—783 (1966).

GINSBERG, H. S., J. F. WILLIAMS, W. DOERFLER, and H. SHIMOJO: Proposed nomenclature for mutants of adenoviruses. J. Virol. **12**, 663—664 (1973).

GINSBERG, H. S., G. F. BADGER, J. H. DINGLE, W. S. JORDAN, JR., and S. KATZ: Etiologic relationship of the RI-67 agent to "Acute Respiratory Disease (ARD)". J. clin. Invest. **34**, 820—831 (1955).

GIORNO, R., and J. R. KATES: Mechanism of inhibition of vaccinia virus replication in adenovirus-infected HeLa cells. J. Virol. **7**, 208—213 (1971).

GIRARDI, A. J., M. R. HILLEMAN, and R. E. ZWICKEY: Tests in hamsters for oncogenic quality of ordinary viruses including adenovirus type 7. Proc. Soc. exp. Biol. (N.Y.) **115**, 1141—1150 (1964).

GOODHEART, C.: DNA density of oncogenic and non-oncogenic simian adenoviruses. Virology **44**, 645—648 (1971).

GRAHAM, F. L., and A. J. VAN DER EB: A new technique for the assay of infectivity of human adenovirus 5 DNA. Virology **52**, 456—467 (1973 a).

GRAHAM, F. L., and A. J. VAN DER EB: Transformation of rat cells by DNA of human adenovirus 5. Virology **54**, 536—539 (1973 b).

GRAHAM, F. L., P. J. ABRAHAMS, S. O. WARNAAR, C. MULDER, F. A. J. DE VRIES, W. FIERS, and A. J. VAN DER EB: Studies on *in vitro* transformation with viral DNA and DNA fragments. Cold Spr. Harb. Symp. quant. Biol. **39**, 637—650 (1974).

GREEN, M.: Biochemical studies on adenovirus multiplication. III. Requirement for DNA synthesis. Virology **18**, 601—613 (1962 a).

GREEN, M.: Studies on the biosynthesis of viral DNA. IV. Isolation, purification and chemical analysis of adenovirus. Cold Spr. Harb. Symp. quant. Biol. **27**, 219—235 (1962 b).

GREEN, M.: Biosynthetic modifications induced by DNA animal viruses. Ann. Rev. Microbiol. **20**, 189—222 (1966).

GREEN, M.: Oncogenic viruses. Ann. Rev. Biochem. **39**, 701—756 (1970).

GREEN, M.: Molecular basis for the attack on cancer. Proc. nat. Acad. Sci. (Wash.) **69**, 1036—1041 (1972).

GREEN, M., and M. PIÑA: Biochemical studies on the adenovirus multiplication. IV. Isolation, purification and chemical analysis of adenovirus. Virology **20**, 199—207 (1963).

GREEN, M., and M. PIÑA: Biochemical studies on adenovirus multiplication. VI. Properties of highly purified tumorigenic human adenoviruses and their DNA's. Proc. nat. Acad. Sci. (Wash.) **51**, 1251—1259 (1964).

GREEN, M., M. PIÑA, and R. C. KIMES: Biochemical studies on adenovirus multiplication. XII. Plaquing efficiencies of purified human adenoviruses. Virology **31**, 562 to 565 (1967 a).

GREEN, M., M. PIÑA, R. C. KIMES, P. C. WENSINK, L. A. MACHATTIE, and C. A. THOMAS, JR.: Adenovirus DNA: I. Molecular weight and conformation. Proc. nat. Acad. Sci. (Wash.) **57**, 1302—1309 (1967 b).

GREEN, M., J. T. PARSONS, M. PIÑA, K. FUJINAGA, H. CAFFIER, and I. LANDGRAF-LEURS: Transcription of adenovirus genes in productively infected and in transformed cells. Cold Spr. Harb. Symp. quant. Biol. **35**, 803—818 (1970).

GREENBERG, J. R.: High stability of messenger RNA in growing cultured cells. Nature (Lond.) **240**, 102—104 (1972).

GRODZICKER, T., J. F. WILLIAMS, P. A. SHARP, and J. SAMBROOK: Physical mapping of cross over events between adenovirus 5 and adenovirus 2. Cold Spr. Harb. Symp. quant. Biol. **39**, 439—446 (1974).

GUTEKUNST, R. R., R. J. WHITE, W. P. EDMONDSON, and R. M. CHANOCK: Immunization with live type 4 adenovirus: Determination of infectious virus dose and protective effect of enteric infection. Amer. J. Epidem. **86**, 341—349 (1967).

HAAS, M., M. VOGT, and R. DULBECCO: Loss of SV40 DNA-RNA hybrids from nitrocellulose membranes. Implications for the study of virus-host DNA interactions. Proc. nat. Acad. Sci. (Wash.) **69**, 2160—2164 (1972).

HARTLEY, J. W., and W. P. ROWE: A new mouse virus apparently related to the adenovirus group. Virology **11**, 645—647 (1960).
HARTLEY, J. W., R. J. HUEBNER, and W. P. ROWE: Serial propagation of adenovirus (APC) in monkey kidney tissue cultures. Proc.Soc.exp.Biol.(N.Y.) **92**, 667—669(1956).
HARUNA, J., H. YAOSI, R. KONO, and I. WATANABE: Separation of adenovirus by chromatography on DEAE-cellulose. Virology **13**, 254—267 (1961).
HASHIMOTO, K., T. SUGIYAMA, and S. SASAKI: An adenovirus isolated from the feces of mice. I. Isolation and identification. Jap. J. Microbiol. **10**, 115—125 (1966).
HASHIMOTO, K., K. NAKAJIMA, K.-I. ODA, and H. SHIMOJO: Complementation of translational defect for growth of human adenovirus type 2 in simian cells by a simian virus 40-induced factor. J. molec. Biol. **81**, 207—223 (1973).
HATCH, G. G., G. L. VAN HOOSIER, K. J. MCCORMICK, and J. J. TRENTIN: Effect of age and virus dose on the susceptibility of inbred hamsters (LSH/LAK) to tumor induction by simian adenovirus type 7 (SA-7). Fed. Proc. **29**, 371 (1970).
HAYASHI, K., and W. C. RUSSELL: A study of the development of adenovirus antigens by the immunofluorescent technique. Virology **34**, 470—480 (1968).
HENRY, P. H., L. E. SCHNIPPER, R. J. SAMAHA, C. S. CRUMPACKER, A. M. LEWIS, JR., and A. S. LEVINE: Studies of nondefective adenovirus 2-simian virus 40 hybrid viruses. VI. Characterization of the DNA from five nondefective hybrid viruses. J. Virol. **11**, 665—671 (1973).
HENSHAW, E. C.: Messenger RNA in rat liver polyribosomes: Evidence that it exists as ribonucleoprotein particles. J. molec. Biol. **36**, 401—411 (1968).
HILLEMAN, M. R.: Adenoviruses, history and future of a vaccine. In: Viruses Inducing Cancer (W. J. BURDETTE, ed.), pp. 377—402. Univ. of Utah Press, 1966.
HILLEMAN, M. R., and J. H. WERNER: Recovery of new agent from patients with acute respiratory illness. Proc. Soc. exp. Biol. (N.Y.) **85**, 183—188 (1954).
HILLEMAN, R. J., J. H. WERNER, and M. T. STEWART: Grouping and occurrence of RI (prototype RI-67) viruses. Proc. Soc. exp. Biol. (N.Y.) **90**, 555—562 (1955).
HILLIS, W. D., and R. GOODMAN: Serological classification of chimpanzee adenoviruses by hemagglutination and hemagglutination inhibition. J. Immunol. **102**, 1089—1095 (1969).
HODGE, L. D., and M. D. SCHARFF: Effect of adenovirus on host cell DNA synthesis in synchronized cells. Virology **37**, 554—564 (1969).
HOGGAN, M. D., W. P. ROWE, P. H. BLACK, and R. J. HUEBNER: Production of "tumor-specific" antigens by oncogenic viruses during acute cytolytic infections. Proc. nat. Acad. Sci. (Wash.) **53**, 12—19 (1965).
HOLLAND, J. J., and E. D. KIEHN: Specific cleavage of viral proteins as steps in the synthesis and maturation of enteroviruses. Proc. nat. Acad. Sci. (Wash.) **60**, 1015 to 1022 (1968).
HOLLINSHEAD, A. C., B. BUNNAG, and T. C. ALFORD: Relationship between the subunits and "T" antigens of adenovirus type 12. Nature (Lond.) **215**, 397—399 (1967).
HOLMES, D. S., and J. BONNER: Preparation, molecular weight, base composition, and secondary structure of giant nuclear ribonucleic acid. Biochemistry **12**, 2330 to 2338 (1973).
HOMBURGER, F., L. P. MERKOW, and M. SLIFKIN: Oncogenic Adenoviruses. Progr. exp. Tumor Res. Vol. 18. Basel: S. Karger 1973.
HORNE, R. W., S. BRENNER, A. P. WATERSON, and P. WILDY: The icosahedral form of an adenovirus. J. molec. Biol. **1**, 84—86 (1959).
HORWITZ, M.: Intermediates in the synthesis of type 2 adenovirus deoxyribonucleic acid. J. Virol. **8**, 675—683 (1971).
HORWITZ, M. S., M. D. SCHARFF, and J. V. MAIZEL, JR.: Synthesis and assembly of adenovirus 2: I. Polypeptide synthesis, assembly of capsomers and morphogenesis of the virion. Virology **39**, 682—694 (1969).
HORWITZ, M. S., J. V. MAIZEL, and M. D. SCHARFF: Molecular weight of adenovirus type 2 hexon polypeptide. J. Virol. **6**, 569—571 (1970).
HORWITZ, M. S., C. BRAYTON, and S. G. BAUM: Synthesis of type 2 adenovirus DNA in the presence of cycloheximide. J. Virol. **11**, 544—551 (1973).

Huebner, R. J.: 70 newly recognized viruses in man. Publ. Hlth Rep. (Wash.) **74**, 6—12 (1959).

Huebner, R. J.: Adenovirus-directed tumor and T antigens. In: Perspect. Virol. (M. Pollard, ed.), Vol. V, pp. 147—166. New York: Academic Press, 1967.

Huebner, R. J., W. P. Rowe, and W. T. Lane: Oncogenic effects in hamsters of human adenovirus type 12 and 18. Proc. nat. Acad. Sci. (Wash.) **48**, 2051—2058 (1962).

Huebner, R. J., W. P. Rowe, H. C. Turner, and W. T. Lane: Specific adenovirus complement-fixing antigens in virus-free hamster and rat tumors. Proc. nat. Acad. Sci. (Wash.) **50**, 379—389 (1963).

Huebner, R. J., M. J. Casey, R. M. Chanock, and K. Schell: Tumors induced in hamsters by a strain of adenovirus type 3: Sharing of tumor antigens and "neoantigens" with those produced by adenovirus type 7 tumors. Proc. nat. Acad. Sci. (Wash.) **54**, 381—388 (1965).

Huebner, R. J., W. P. Rowe, T. G. Ward, R. H. Parrot, and J. A. Bell: Adenoidal-pharyngeal-conjunctival agents. A newly recognized group of common viruses of the respiratory system. New Engl. J. Med. **251**, 1077—1086 (1954).

Huebner, R. J., H. G. Pereira, A. C. Allison, A. C. Hollinshead, and H. C. Turner: Production of type-specific C antigen in virus-free hamster tumor cells induced by adenovirus type 12. Proc. nat. Acad. Sci. (Wash.) **51**, 432—439 (1964).

Hull, R. N., I. S. Johnson, C. G. Culbertson, C. B. Reiner, and H. F. Wright: Oncogenicity of the simian adenoviruses. Science **150**, 1044—1046 (1965).

Ishibashi, M.: Retention of viral antigen in the cytoplasm of cells infected with temperature-sensitive mutants of an avian adenovirus. Proc. nat. Acad. Sci. (Wash.) **65**, 304—309 (1970).

Ishibashi, M.: Temperature-sensitive conditional lethal mutants of an avian adenovirus (CELO). I. Isolation and characterization. Virology **45**, 42—52 (1971).

Ishibashi, M., and J. V. Maizel, Jr.: The polypeptides of adenovirus. V. Young virions, structural intermediate between top components and aged virions. Virology **57**, 409—424 (1974 a).

Ishibashi, M., and J. V. Maizel, Jr.: The polypeptides of adenovirus. VI. Early and late glycopolypeptides. Virology **58**, 345—361 (1974 b).

Jacobson, M. F., and D. Baltimore: Polypeptide cleavages in the formation of poliovirus proteins. Proc. nat. Acad. Sci. (Wash.) **61**, 77—84 (1968).

Jawetz, E., P. Thygeson, L. Hanna, A. Nicholas, and S. Kumura: The etiology of epidemic keratoconjunctivitis. Amer. J. Ophthal. **43**, 79—83 (1957).

Jelinek, W., M. Adesnik, M. Salditt, D. Sheiness, R. Wall, G. Molloy, L. Philipson, and J. E. Darnell: Further evidence on the nuclear origin and transfer to the cytoplasm of poly (A) sequences in mammalian cell RNA. J. molec. Biol. **73**, 515—532 (1973).

Jones, R. F., B. B. Asch, and D. S. Yohn: On the oncogenic properties of chicken embryo lethal orphan virus, an avian adenovirus. Cancer Res. **30**, 1580—1585 (1970).

Jörnvall, H., H. Ohlsson, and L. Philipson: An acetylated N-terminus of adenovirus type 2 hexon protein. Biochem. biophys. Res. Commun. **56**, 304—310 (1974 a).

Jörnvall, H., U. Pettersson, and L. Philipson: Structural studies of adenovirus type 2 hexon protein. Europ. J. Biochem. **48**, 179—192 (1974 b).

Kapsenberg, J. G.: Relationship of infectious canine hepatitis virus to human adenovirus. Proc. Soc. exp. Biol. (N.Y.) **101**, 611—614 (1959).

Kasel, J. A., and M. Huber: Relationship of the adenovirus erythrocyte-receptor modifying factor to the type specific complement fixing antigen. Proc. Soc. exp. Biol. (N.Y.) **116**, 16—18 (1964).

Kasel, J. A., W. P. Rowe, and J. L. Nemes: Further characterization of the adenovirus erythrocyte-receptor modifying factor. J. exp. Med. **114**, 717—728 (1961).

Kasel, J. A., F. Lode, and V. Knight: Infection of volunteers with adenovirus type 16. Proc. Soc. exp. Biol. (N.Y.) **114**, 621—623 (1963 a).

Kasel, J. A., H. E. Evans, A. Spickard, and V. Knight: Conjunctivitis and enteric infection with adenovirus types 26 and 27: Responses to primary, secondary and reciprocal cross-challenges. Amer. J. Hyg. **77**, 265—282 (1963 b).

KASEL, J. A., M. HUBER, F. LODA, P. A. BANKS, and V. KNIGHT: Immunization of volunteers with soluble antigens of adenovirus type 1. Proc. Soc. exp. Biol. (N.Y.) **117**, 186—190 (1964).

KASEL, J. A. R. H. ALFORD, J. R. LEHRICH, P. A. BANKS, M. HUBER, and V. KNIGHT: Adenovirus soluble antigens for human immunization. A progress report. Amer. Rev. resp. Dis. **94**, 168—174 (1966).

KASZA, L.: Isolation of an adenovirus from the brain of a pig. Amer. J. vet. Res. **27**, 751—758 (1966).

KATES, J.: Transcription of the vaccinia virus genome and the occurrence of polyriboadenylic acid sequences in messenger RNA. Cold Spr. Harb. Symp. quant. Biol. **35**, 743—752 (1970).

KATZ, S., G. F. BADGER, A. B. DENISON, F. W. DENNY, JR., and W. S. JORDAN, JR.: Search for illness due to adenovirus type 4 among college dormitory freshmen. Proc. Soc. exp. Biol. (N.Y.) **101**, 592—594 (1959).

KAWAMURA, H., F. SHIMIZU, and H. TSUBAHARA: Avian adenovirus: Its properties and serological classification. Nat. Inst. Anim. Hlth Quart. (Tokyo) **4**, 183—193 (1964).

KELLY, T., and J. ROSE: Simian virus 40 integration site in an adenovirus 7-SV40 hybrid DNA molecule. Proc. nat. Acad. Sci. (Wash.) **68**, 1037—1041 (1971).

KELLY, T., and A. M. LEWIS, JR.: Use of non-defective adenovirus-simian virus 40 hybrids for mapping the simian virus 40. J. Virol. **12**, 643—652 (1973).

KIM, C. S., E. A. SUELTENFUSS, and S. S. KALTER: Isolation and characterization of simian adenoviruses isolated in association with an outbreak of pneumoenteritis in vervet monkeys *(Cercopithecus aethiops)*. J. infect. Dis. **117**, 292—300 (1967).

KIMES, R., and M. GREEN: Adenovirus DNA. II. Separation of molecular halves of adenovirus type 2. J. molec. Biol. **50**, 203—205 (1970).

KIT, S., L. J. PIEKARSKI, D. R. DUBBS, R. A. DE TORRES, and M. ANKEN: Enzyme induction in green monkey kidney cultures infected with simian adenovirus. J. Virol. **1**, 10—15 (1967).

KJELLÉN, L., and H. G. PEREIRA: Role of adenovirus antigens in the induction of virus neutralizing antibody. J. gen. Virol. **2**, 177—185 (1968).

KJELLÉN, L., G. LAGERMALM, A. SVEDMYR, and K. G. THORSSON: Crystalline-like patterns in the nuclei of cells infected with an animal virus. Nature (Lond.) **175**, 505—506 (1955).

KLEMPERER, H. G., and H. G. PEREIRA: Study of adenovirus antigens fractionated by chromatography on DEAE-cellulose. Virology **9**, 536—545 (1959).

KLINE, L. K., S. M. WEISSMAN, and D. SOLL: Investigation on adenovirus-directed 4S RNA. Virology **48**, 291—296 (1972).

KOCZOT, F. J., B. J. CARTER, C. F. GARON, and J. A. ROSE: Self-complementarity of terminal sequences within plus or minus strands of adenovirus-associated virus DNA. Proc. nat. Acad. Sci. (Wash.) **70**, 215—219 (1973).

KÖHLER, K.: Reinigung und Charakterisierung zweier Proteine des Adenovirus Typ 2. Z. Naturforsch. **20 b**, 747—752 (1965).

KUBINSKI, H., and J. A. ROSE: Regions containing repeating base-pairs in DNA from some oncogenic and non-oncogenic animal viruses. Proc. nat. Acad. Sci. (Wash.) **57**, 1720—1725 (1967).

KUMAR, A., and U. LINDBERG: Characterization of messenger ribonucleoprotein and messenger RNA from KB cells. Proc. nat. Acad. Sci. (Wash.) **69**, 681—685 (1972).

KUSANO, T., and I. YAMANE: Transformation *in vitro* of the embryonal hamster brain cells by human adenovirus type 12. Tohoku J. exp. Med. **92**, 141—150 (1967).

LACY, S., SR., and M. GREEN: Biochemical studies on adenovirus multiplication. VII. Homology between DNA's of tumorigenic and nontumorigenic human adenoviruses. Proc. nat. Acad. Sci. (Wash.) **52**, 1053—1059 (1964).

LACY, S., SR., and M. GREEN: Adenovirus multiplication. Genetic relatedness of tumorigenic human adenovirus type 7, 12, 18. Science **150**, 1296—1298 (1965).

LACY, S., SR., and M. GREEN: The mechanism of viral carcinogenesis by DNA mammalian viruses: DNA-DNA homology relationship among the "weakly" oncogenic human adenoviruses. J. gen. Virol. **1**, 413—418 (1967).

LANDGRAF-LEURS, M., and M. GREEN: Adenovirus DNA. III. Separation of the complementary strands of adenovirus types 2, 7 and 12 DNA molecules. J. molec. Biol. **60**, 185—202 (1971).

LANDGRAF-LEURS, M., and M. GREEN: DNA strand selection during the transcription of the adenovirus 2 genome in infected and transformed cells. Biochim. biophys. Acta (Amst.) **312**, 667—673 (1973).

LARSON, V. M., A. J. GIRARDI, M. R. HILLEMAN, and R. E. ZWICKEY: Studies on oncogenicity of adenovirus type 7 viruses in hamsters. Proc. Soc. exp. Biol. (N.Y.) **118**, 15—24 (1965).

LAVER, W. G.: Isolation of an arginine-rich protein from particles of adenovirus type 2. Virology **41**, 488—500 (1970).

LAVER, W. G., J. R. SURIANO, and M. GREEN: Adenovirus proteins. II. N-terminal amino acid analysis. J. Virol. **1**, 723—728 (1967).

LAVER, W. G., N. G. WRIGLEY, and H. G. PEREIRA: Removal of pentons from particles of adenovirus type 2. Virology **38**, 599—605 (1969).

LAVER, W. G., H. B. YOUNGHUSBAND, and N. G. WRIGLEY: Purification and properties of chick embryo lethal orphan virus (an avian adenovirus). Virology **45**, 598—614 (1971).

LAVER, W. G., H. G. PEREIRA, W. C. RUSSELL, and R. C. VALENTINE: Isolation of an internal component from adenovirus type 5. J. molec. Biol. **37**, 379—386 (1968).

LAWRENCE, W. C., and H. S. GINSBERG: Intracellular uncoating of type 5 adenovirus deoxyribonucleic acid. J. Virol. **1**, 851—867 (1967).

LEBLEU, B., G. MARBAIX, G. HUEZ, J. TEMMERMAN, A. BURNY, and H. CHANTRENNE: Characterization of the messenger ribonucleoprotein released from reticulocyte polyribosomes by EDTA treatment. Europ. J. Biochem. **19**, 264—269 (1971).

LEDINKO, N.: Stimulation of DNA synthesis and thymidine kinase activity in human embryonic kidney cells infected by adenovirus 2 or 12. Cancer Res. **27**, 1459—1469 (1967).

LEDINKO, N.: Transient stimulation of deoxyribonucleic acid-dependent ribonucleic acid polymerase and histone acetylation in human embryonic kidney cultures infected with adenovirus 2 or 12: Apparent induction of host ribonucleic acid synthesis. J. Virol. **6**, 58—68 (1970).

LEDINKO, N.: Inhibition by α-aminitin of adenovirus 12 replication in human embryonic kidney cells and of adenovirus transformation of hamster cells. Nature (New Biol.) **233**, 247—248 (1971).

LEDINKO, N.: Nucleolar ribosomal precursor RNA and protein metabolism in human embryo kidney cultures infected with adenovirus 12. Virology **49**, 79—89 (1972).

LEE, Y., J. MENDECKI, and G. BRAWERMAN: A polynucleotide segment rich in adenylic acid in the rapidly labelled polyribosomal RNA component of mouse sarcoma 180 ascites cells. Proc. nat. Acad. Sci. (Wash.) **68**, 1331—1335 (1971).

LEIJONMARCK, M., O. RÖNNQUIST, P.-E. WERNER, and L. PHILIPSON: On the quarternary structure of adenovirus type 2 hexons and their arrangement in the cubic crystal cell. Acta crystallographica (submitted, 1974).

LEVINE, A. J., and H. S. GINSBERG: Mechanism by which fiber antigen inhibits multiplication of type 5 adenovirus. J. Virol. **1**, 747—757 (1967).

LEVINE, A. J., and H. S. GINSBERG: Role of adenovirus structural proteins in the cessation of host-cell biosynthetic functions. J. Virol. **2**, 430—439 (1968).

LEVINE, A. S., M. J. LEVIN, M. N. OXMAN, and A. M. LEWIS: Studies of nondefective adenovirus 2-simian virus 40 hybrid viruses. VII. Characterization of the simian virus 40 RNA species induced by five nondefective hybrid viruses. J. Virol. **11**, 672—681 (1973).

LEVINTHAL, J. D., and W. PETERSON: *In vitro* transformation and immunofluorescence with human adenovirus type 12 in rat and rabbit kidney cells. Fed. Proc. **24**, 174 (1965).

LEWIS, A. M., JR., and W. P. ROWE: Isolation of two plaque variants from the adenovirus type 2-simian virus 40 hybrid population which differ in their efficiency in yielding simian virus 40. J. Virol. **5**, 413—420 (1970).

Lewis, A. M., Jr., and W. P. Rowe: Studies on nondefective adenovirus-simian virus 40 hybrid viruses. I. A newly characterized simian virus 40 antigen induced by the Ad2+ND$_1$ virus. J. Virol. **7**, 189—197 (1971).
Lewis, A. M., Jr., W. H. Wiese, and W. P. Rowe: The presence of antibodies in human serum to early (T) adenovirus antigens. Proc. nat. Acad. Sci. (Wash.) **57**, 622—629 (1967).
Lewis, A. M., Jr., A. S. Rabson, and A. S. Levine: Studies on nondefective adenovirus 2-simian virus 40 hybrid viruses. X. Transformation of hamster kidney cells by adenovirus 2 and the nondefective hybrid viruses. J. Virol. **14**, 1290—1301 (1974).
Lewis, A. M., Jr., S. G. Baum, K. O. Prigge, and W. P. Rowe: Occurrence of adenovirus-SV40 hybrids among monkey kidney cell-adapted strains of adenovirus. Proc. Soc. exp. Biol. (N.Y.) **122**, 214—218 (1966).
Lewis, A. M., Jr., M. J. Levin, W. H. Wiese, C. S. Crumpacker, and P. H. Henry: A non-defective (competent) adenovirus-SV40 hybrid isolated from the Ad2-SV40 hybrid population. Proc. nat. Acad. Sci. (Wash.) **63**, 1128—1135 (1969).
Lewis, A. M., Jr., A. S. Levine, C. S. Crumpacker, M. J. Levin, R. J. Samaha, and P. H. Henry: Studies of nondefective adenovirus 2-simian virus 40 hybrid viruses. V. Isolation of additional hybrids which differ in their simian virus 40 specific biological properties. J. Virol. **11**, 655—664 (1973).
Liem, I. T. F., I. Kron und R. Wigand: Dodecon, das lösliche Hämagglutinin bei allen Adenoviren der Untergruppe II. Arch. ges. Virusforsch. **33**, 177—181 (1971).
Lindberg, U., and J. E. Darnell: SV40 specific RNA in the nucleus and polyribosomes of transformed cells. Proc. nat. Acad. Sci. (Wash.) **65**, 1089—1096 (1970).
Lindberg, U., and T. Persson: Isolation of mRNA from KB-cells by affinity chromatography on polyuridylic acid covalently linked to Sepharose. Europ. J. Biochem. **31**, 246—254 (1972).
Lindberg, U., and B. Sundquist: Isolation of messenger ribonucleoproteins from mammalian cells. J. molec. Biol. **86**, 451—468, (1974).
Lindberg, U., T. Persson, and L. Philipson: Isolation and characterization of adenovirus messenger RNA in productive infection. J. Virol. **10**, 909—919 (1972).
Lonberg-Holm, K., and L. Philipson: Early events of virus-cell interaction in an adenovirus system. J. Virol. **4**, 323—338 (1969).
Loni, M. C., and M. Green: Detection of viral DNA sequences in adenovirus transformed cells by *in situ* hybridization. J. Virol. **12**, 1288—1292 (1973).
López-Revilla, R. and G. Walter: Polypeptide specific for cells with adenovirus 2-SV40 hybrid Ad2+ND$_1$. Nature (New Biol.) **244**, 165—167 (1973).
Lucas, J. J., and H. S. Ginsberg: Synthesis of virus-specific ribonucleic acid in KB cells infected with type 2 adenovirus. J. Virol. **8**, 203—213 (1971).
Lucas, J. J., and H. S. Ginsberg: Transcription and transport of virus-specific ribonucleic acids in African green monkey kidney cells abortively infected with type 2 adenovirus. J. Virol. **10**, 1109—1118 (1972 a).
Lucas, J. J., and H. S. Ginsberg: Identification of double-stranded virus-specific ribonucleic acid in KB cells infected with type 2 adenovirus Biochem. biophys. Res. Commun. **49**, 39—44 (1972 b).
Lundholm, U., and W. Doerfler: Temperature-sensitive mutants of adenovirus type 12. Virology **45**, 827—829 (1971).
Lwoff, A., R. Horne, and P. Tournier: A system of viruses. Cold Spr. Harb. Symp. quant. Biol. **27**, 51—55 (1962).
Maden, B. E. H., M. H. Vaughan, J. R. Warner, and J. E. Darnell: Effects of valine deprivation on ribosome formation in HeLa cells. J. molec. Biol. **45**, 265—275 (1969).
Magnusson, G., V. Pigiet, E.-L. Winnacker, R. Abrams, and P. Reichard: RNA-linked short DNA fragments during polyoma replication. Proc. nat. Acad. Sci. (Wash.) **70**, 412—415 (1973).
Maizel, J. V., Jr.: Polyacrylamide gel electrophoresis of viral proteins. In: Methods in Virology, Vol. V. (K. Maramorosch and H. Koprowski, eds.), pp. 179—246. New York: Academic Press, 1971.

MAIZEL, J. V., JR., D. O. WHITE, and M. D. SCHARFF: The polypeptides of adenovirus. I. Evidence of multiple protein components in the virion and a comparison of types 2, 7 and 12. Virology **36**, 115—125 (1968 a).

MAIZEL, J. V., JR., D. O. WHITE, and M. D. SCHARFF: The polypeptides of adenovirus. II. Soluble proteins, cores, top components and the structure of the virion. Virology **36**, 126—136 (1968 b).

MAK, S.: Defective virions in human adenovirus type 12. J. Virol. **7**, 426—433 (1971).

MAK, S., and M. GREEN: Biochemical studies on adenovirus multiplication. XIII. Synthesis of virus-specific ribonucleic acid during infection with human adenovirus type 12. J. Virol. **2**, 1055—1063 (1968).

MARTINÉZ-PALOMO, A., and C. BRAILOVSKY: Surface layer in tumor cells transformed by adeno-12 and SV 40 viruses. Virology **34**, 379—382 (1968).

MARUSYK, R., E. NORRBY, and U. LUNDQUIST: Biophysical comparison of two canine adenoviruses. J. Virol. **5**, 507—512 (1970).

MARUSYK, R., E. NORRBY, and H. MARUSYK: The relationship of adenovirus-induced paracrystalline structures to the virus core protein(s). J. gen. Virol. **14**, 261—270 (1972).

MAUTNER, V., and H. G. PEREIRA: Crystallization of a second adenovirus protein (the fiber). Nature (Lond.) **230**, 456—457 (1971).

MAYNE, N., J. P. BURNETT, and L. K. BUTLER: Tumour induction by simian adenovirus SA 7 DNA fragments. Nature (New Biol.) **232**, 182—183 (1971).

McALLISTER, R. M., and I. MACPHERSON: Transformation of a hamster cell line by adenovirus type 12. J. gen. Virol. **2**, 99—106 (1968).

McALLISTER, R. M., B. H. LANDING, and C. R. GOODHEART: Isolation of adenoviruses from neoplastic and non-neoplastic tissues of children. Lab. Invest. **13**, 894—901 (1964).

McALLISTER, R. M., R. V. GILDEN, and M. GREEN: Adenoviruses in human cancer. Lancet **i**, 831—833 (1972).

McALLISTER, R. M., C. R. GOODHEART, V. Q. MIRABAL, and R. J. HUEBNER: Human adenoviruses: Tumor production in hamsters by types 12 and 18 grown from single plaques. Proc. Soc. exp. Biol. (N.Y.) **122**, 455—458 (1966).

McALLISTER, R. M., M. O. NICOLSON, A. M. LEWIS, JR., I. MACPHERSON, and R. J. HUEBNER: Transformation of rat embryo cells by adenovirus type 1. J. gen. Virol. **4**, 29—36 (1969 a).

McALLISTER, R. M., M. O. NICOLSON, G. REED, J. KERN, R. V. GILDEN, and R. J. HUEBNER: Transformation of rodent cells by adenovirus 19 and other group D adenoviruses. J. nat. Cancer Inst. **43**, 917—923 (1969 b).

McBRIDE, W. D., and A. WIENER: *In vitro* transformation of hamster kidney cells by human adenovirus type 12. Proc. Soc. exp. Biol. (N.Y.) **115**, 870—874 (1964).

McCHESNEY, A. E., J. J. ENGLAND, and L. J. RICH: Adenoviral infections in foals. J. Amer. vet. med. Ass. **162**, 545—549 (1974).

McDOUGALL, J. K., A. R. DUNN, and K. JONES: *In situ* hybridization of adenovirus RNA and DNA. Nature (Lond.) **236**, 346—348 (1972).

McDOUGALL, J. K., R. KUCHERLAPATI, and F. H. RUDDLE: Localization and induction of the human thymidine kinase gene by adenovirus 12. Nature (New Biol.) **245**, 172—175 (1973).

McDOUGALL, J. K., A. R. DUNN, and P. H. GALLIMORE: Recent studies on the characteristics of adenovirus infected and transformed cells. Cold Spr. Harb. Symp. quant. Biol. **39**, 591—600 (1974).

McFERRAN, J. B., J. K. CLARKE, and T. J. CONNOR: Serological classification of avian adenoviruses. Arch. ges. Virusforsch. **39**, 132—139 (1972).

McFERRAN, J. B., R. NELSON, J. M. McCRACKEN, and J. G. ROSS: Viruses isolated from sheep. Nature (Lond.) **221**, 194—195 (1969).

McGUIRE, P. M., C. SWART, and L. D. HODGE: Adenovirus messenger RNA in mammalian cells: Failure of polyribosome association in the absence of nuclear cleavage. Proc. nat. Acad. Sci. (Wash.) **69**, 1578—1582 (1972).

McIntosh, K., S. Payne, and W. C. Russel: Studies on lipid metabolism in cells infected with adenovirus. J. gen. Virol. **10**, 251—265 (1971).

Mendecki, J., S. Lee, and C. Brawerman: Characteristics of the polyadenylic acid segment associated with messenger ribonucleic acid in mouse sarcoma 180 ascites cells. Biochemistry **11**, 792—798 (1972).

Missal, O.: Isolierung von Adenovirus aus einer Mäusekolonie und dessen Charakterisierung. Arch. exp. Vet.-Med. **23**, 783—791 (1969).

Mohanty, S. B.: Comparative study of bovine adenoviruses. Amer. J. vet. Res. **32**, 1899—1905 (1971).

Molloy, G. R., M. B. Sporn, D. E. Kelley, and R. P. Perry: Localization of polyadenylic acid sequences in messenger ribonucleic acid of mammalian cells. Biochemistry **11**, 3256—3260 (1972).

Morel, C., B. Kayibanda, and K. Scherrer: Proteins associated with globin messenger RNA in avian erythroblasts: Isolation and comparison with the proteins bound to nuclear messenger-like RNA. FEBS Letters **18**, 84—88 (1971).

Morel, C., E. S. Gander, M. Herzberg, J. Dubrochet, and K. Scherrer: The duck-globin messenger-ribonucleoprotein complex. Resistance to high ionic strength, particle gel electrophoresis, composition and visualisation by dark-field electron microscopy. Europ. J. Biochem. **36**, 455—464 (1973).

Morgan, C., H. S. Rosenkranz, and B. Mednis: Structure and development of viruses as observed in the electron microscope. X. Entry and uncoating of adenoviruses. J. Virol. **4**, 777—796 (1969).

Morgan, C., G. Goodman, H. Rose, C. Howe, and J. Huang: Electron microscopic and histochemical studies of an unusual crystalline protein occurring in cells infected by type 5 adenovirus. J. biophys. biochem. Cytol. **3**, 505—508 (1957).

Morrow, J. F., and P. Berg: Cleavage of simian virus 40 DNA at a unique site by a bacterial restriction enzyme. Proc. nat. Acad. Sci. (Wash.) **69**, 3365—3369 (1972).

Mulder, C., P. A. Sharp, H. Delius, and U. Pettersson: Specific fragmentation of DNA of adenovirus serotypes 3, 5, 7 and 12, and adeno-SV40 hybrid virus Ad2$^+$ ND$_1$ by restriction endonuclease *Eco* RI. J. Virol. **14**, 68—77 (1974 a).

Mulder, C., J. R. Arrand, H. Delius, W. Keller, U. Pettersson, R. J. Roberts, and P. A. Sharp: Cleavage maps from adenovirus types 2 and 5 by restriction endonuclease *Eco*RI and *Hpa*I. Cold Spr. Harb. Symp. quant. Biol. **39**, 397—400 (1974 b).

Murray, R. E., and M. Green: Adenovirus DNA. IV. Topology of adenovirus genomes. J. molec. Biol. **74**, 735—738 (1973).

Naegele, R. F., and F. Rapp: Enhancement of the replication of human adenoviruses in simian cells by simian adenovirus SV 15. J. Virol. **1**, 838—840 (1967).

Neurath, A. R., B. A. Rubin, and J. T. Stasny: Cleavage by formamide of intercapsomer bonds in adenovirus type 4 and 7 virions and hemagglutinins. J. Virol. **2**, 1086—1095 (1968).

Neurath, A. R., R. W. Hartzell, and B. A. Rubin: Solubilization and some properties of erythrocyte receptors for adenovirus type 7 hemagglutinin. Nature (Lond.) **221**, 1069—1071 (1969).

Neurath, A. R., B. A. Rubin, F. P. Wiener, and R. W. Hartzell: Evidence of the structure and function of virion proteins of adenoviruses. Presence of half-cystine residues in the capsid of adenovirus type 7. FEBS Letters **7**, 114—118 (1970 a).

Neurath, A. R., J. T. Stasny, and B. A. Rubin: Disruption of adenovirus type 7 by lithium iodide resulting in the release of viral deoxyribonucleic acid. J. Virol. **5**, 173—178 (1970 b).

Nicolson, M. O., and R. M. McAllister: Infectivity of human adenovirus 1 DNA. Virology **48**, 14—21 (1972).

Norrby, E.: The relationship between the soluble antigens and the virion of adenovirus type 3. I. Morphological characteristics. Virology **28**, 236—248 (1966 a).

Norrby, E.: The relationship between the soluble antigens and the virion of type 3. II. Identification and characterization of an incomplete hemagglutinin. Virology **30**, 608—617 (1966 b).

Norrby, E.: Biological significance of structural adenovirus components. Curr. Top. Microbiol. Immunol. **43**, 1—43 (1968).

Norrby, E.: The relationship between soluble antigens and the virion of adenovirus type 3. IV. Immunological characteristics. Virology **37**, 565—576 (1969 a).

Norrby, E.: The structural and functional diversity of adenovirus capsid components. J. gen. Virol. **5**, 221—236 (1969 b).

Norrby, E.: Capsid mosaic of intermediate strains of human adenoviruses. J. Virol. **4**, 657—662 (1969 c).

Norrby, E.: Adenoviruses. In: Comparative Virology (K. Maramorosch and E. Kurstak, eds.), pp. 105—134. New York: Academic Press, 1971.

Norrby, E., and P. Skaaret: The relationship between soluble antigens and the virion of adenovirus type 3. III. Immunological identification of fiber antigen and isolated vertex capsomer antigen. Virology **32**, 489—502 (1967).

Norrby, E., and J. Ankerst: Biological characterization of structural components of adenovirus type 12. J. gen. Virol. **5**, 183—194 (1969).

Norrby, E., and G. Wadell: Immunological relationship between hexons of certain human adenoviruses. J. Virol. **4**, 663—670 (1969).

Norrby, E., and Y. Gollmar: Mosaics of capsid components produced by cocultivation of certain human adenoviruses *in vitro*. Virology **44**, 383—395 (1971).

Norrby, E., H. Marusyk, and M.-L. Hammarskjöld: The relationship between the soluble antigens and the virion of adenovirus type 3. V. Identification of antigen specificities available at the surface of virions. Virology **38**, 477—482 (1969 a).

Norrby, E., G. Wadell, and H. Marusyk: Fiber-associated incomplete and complete hemagglutinins of adenovirus type 6. Arch. ges. Virusforsch. **28**, 239—249 (1969 b).

Numazaki, Y., S. Shigeta, T. Kumasaka, T. Miyazawa, M. Yamanaka, N. Yano, S. Takai, and N. Ishida: Acute hemorrhagic cystitis in children. Isolation of adenovirus type 11. New Engl. J. Med. **278**, 700—704 (1968).

Öberg, B., J. Saborio, T. Persson, E. Everitt, and L. Philipson: Identification of the *in vitro* translation products of adenovirus mRNA by immunoprecipitation. J. Virol. **15**, 199—207 (1975).

O'Conor, G. T., A. S. Rabson, I. K. Berezesky, and F. J. Paul: Mixed infection with simian virus 40 and adenovirus 12. J. nat. Cancer Inst. **31**, 903—917 (1963).

Ogino, T., and M. Takahashi: Altered properties of thymidine kinase induced in hamster kidney cells by adenovirus type 5 and 12. Biken J. **12**, 17—23 (1969).

Ohe, K.: Virus-coded origin of a low molecular weight RNA from KB cells infected with adenovirus 2. Virology **47**, 726—733 (1972).

Ohe, K., and S. M. Weissman: Nucleotide sequence of an RNA from cells infected with adenovirus type 2. Science **167**, 879—881 (1970).

Ohe, K., S. M. Weissman, and N. R. Cooke: Studies on the origin of a low molecular weight ribonucleic acid from human cells infected with adenoviruses. J. biol. Chem. **244**, 5320—5332 (1969).

Okazaki, R., T. Okazaki, K. Sakabe, K. Sugimoto, and A. Sugino: Mechanism of DNA chain growth. I. Possible discontinuity and unusual secondary structure of newly synthesized chains. Proc. nat. Acad. Sci. (Wash.) **59**, 598—605 (1968).

Okubo, C. K., and H. J. Raskas: Thermosensitive events in the replication of adenovirus type 2 at 42°. Virology **46**, 175—182 (1971).

Okubo, C. K., and H. J. Raskas: A reconstituted system for *in vitro* synthesis of adenovirus 2 proteins. Virology **47**, 487—490 (1972).

Parsons, J. T., and M. Green: Biochemical studies on adenovirus multiplication. XVIII. Resolution of early virus specific RNA species in adeno 2-infected and transformed cells. Virology **45**, 154—162 (1971).

Parsons, J. T., J. Gardner, and M. Green: Studies on adenovirus multiplication. XIX. Resolution of late viral RNA species in the nucleus and cytoplasm. Proc. nat. Acad. Sci. (Wash.) **68**, 557—560 (1971).

Patch, C. T., A. M. Lewis, and A. S. Levine: Evidence for a transcription control region of SV40 in the adenovirus 2-SV40 hybrid Ad2^+ND$_1$. Proc. nat. Acad. Sci. (Wash.) **69**, 3375—3379 (1972).

PATCH, C. T., A. M. LEWIS, JR., and A. S. LEVINE: Studies on nondefective adenovirus 2-simian virus 40 hybrid viruses. IX. Template topography in the early region of simian virus 40. J. Virol. **13**, 677—689 (1974).

PEARSON, G. D., and P. C. HANAWALT: Isolation of DNA replication complexes from uninfected and adenovirus infected HeLa cells. J. molec. Biol. **62**, 65—80 (1971).

PEDERSEN, I. R., and H. S. GINSBERG: Characterization of a new viral component of type 5 adenovirus by immunoelectrophoresis. Acta path. microbiol. scand. **187**, 82—83 (1967).

PEDERSON, T.: Proteins associated with heterogeneous nuclear RNA in eukaryotic cells. J. molec. Biol. **83**, 163—183 (1974).

PEREIRA, H. G.: A protein factor responsible for the early cytopathic effect of adenoviruses. Virology **6**, 601—611 (1958).

PEREIRA, H. G.: A virus inhibitor produced in HeLa cells infected with adenoviruses. Virology **11**, 590—602 (1960).

PEREIRA, H. G., and B. KELLY: Dose response curves of toxic and infective actions of adenovirus in HeLa cell cultures. J. gen. Microbiol. **17**, 517—564 (1957a).

PEREIRA, H. G., and B. KELLY: Latent infection of rabbits by adenovirus type 5. Nature (Lond.) **180**, 615—616 (1957b).

PEREIRA, H. G., and B. KELLY: Studies on natural and experimental infections by adenovirus. Proc. roy. Soc. Med. **50**, 755—757 (1957c).

PEREIRA, H. G., and M. V. T. DE FIGUEIREDO: Mechanism of hemagglutination by adenovirus types 1, 2, 4, 5 and 6. Virology **18**, 1—8 (1962).

PEREIRA, H. G., and W. G. LAVER: Comparison of adenovirus types 2 and 5 hexons by immunological and biochemical techniques. J. gen. Virol. **9**, 163—167 (1970).

PEREIRA, H. G., and J. J. SKEHEL: Spontaneous and tryptic degradation of virus particles and structural components of adenoviruses. J. gen. Virol. **12**, 13—24 (1971).

PEREIRA, H. G., A. C. ALLISON, and C. FARTHING: Study of adenovirus antigens by immunoelectrophoresis. Nature (Lond.) **183**, 895—896 (1959).

PEREIRA, M. S., H. G. PEREIRA, and S. K. CLARKE: Human adenovirus type 31: a new serotype with oncogenic properties. Lancet **i**, 21—23 (1965).

PEREIRA, H. G., R. C. VALENTINE, and W. C. RUSSELL: Crystallization of an adeno virus protein (the hexon). Nature (Lond.) **219**, 946—947 (1968).

PEREIRA, H. G., R. J. HUEBNER, H. S. GINSBERG, and J. VAN DER VEEN: A short description of the adenovirus group. Virology **20**, 613—620 (1963).

PERLMAN, S., M. HIRSCH, and S. PENMAN: Utilization of messenger in adenovirus-2-infected cells at normal and elevated temperatures. Nature (New Biol.) **238**, 143 to 144 (1972).

PERRY, R. P., and D. E. KELLEY: Messenger RNA-protein complexes and newly synthesized ribosomal subunits. Analysis of free particles and components of ribosomes. J. molec. Biol. **35**, 37—59 (1968).

PETTERSSON, U.: Structural proteins of adenoviruses. VI. On the antigenic determinants of the hexon. Virology **43**, 123—136 (1971).

PETTERSSON, U.: Some unusual properties of replicating adenovirus type 2 DNA. J. molec. Biol. **81**, 521—527 (1973).

PETTERSSON, U., and S. HÖGLUND: Structural proteins of adenoviruses. III. Purification and characterization of the adenovirus type 2 penton antigen. Virology **39**, 90—106 (1969).

PETTERSSON, U., and J. SAMBROOK: Amount of viral DNA in the genome of cells transformed by adenovirus type 2. J. molec. Biol. **73**, 125—130 (1973).

PETTERSSON, U., and L. PHILIPSON: Synthesis of complementary RNA sequences during productive adenovirus infection. Proc. nat. Acad. Sci. (Wash.) **71**, 4887—4891 (1974).

PETTERSSON, U., L. PHILIPSON, and S. HÖGLUND: Structural proteins of adenoviruses. I. Purification and characterization of adenovirus type 2 hexon antigen. Virology **33**, 575—590 (1967).

PETTERSSON, U., L. PHILIPSON, and S. HÖGLUND: Structural proteins of adenoviruses. II. Purification and characterization of adenovirus type 2 fiber antigen. Virology **35**, 204—215 (1968).

Pettersson, U., C. Tibbetts, and L. Philipson: manuscript in preparation (1975).
Pettersson, U., C. Mulder, H. Delius, and P. Sharp: Cleavage of adenovirus type 2 DNA into six unique fragments with endonuclease. R·RI. Proc. nat. Acad. Sci. (Wash.) **70**, 200—204 (1973).
Pettersson, U., J. Sambrook, H. Delius, and C. Tibbetts: *In vitro* transcription of adenovirus 2 DNA by *Escherichia coli* RNA polymerase. Virology **59**, 153—167 (1974).
Philipson, L.: Separation on DEAE-cellulose of components associated with adenovirus reproduction. Virology **10**, 459—465 (1960).
Philipson, L.: Attachment and eclipse of adenovirus. J. Virol. **1**, 868—875 (1967).
Philipson, L., and U. Pettersson: Structure and function of virion proteins of adenoviruses. Progr. exp. Tumor Res. (Basel) **18**, 1—55 (1973).
Philipson, L., and U. Lindberg: Reproduction of adenoviruses. In: Comprehensive Virology (H. Fraenkel-Conrat and R. Wagner, eds.), Vol. 3, pp. 143—207. New York-London: Plenum Press, 1974.
Philipson, L., K. Lonberg-Holm, and U. Pettersson: Virus receptor interaction in an adenovirus system. J. Virol. **2**, 1064—1075 (1968).
Philipson, L., R. Wall, G. Glickman, and J. E. Darnell: Addition of polyadenylate sequences to virus-specific RNA during adenovirus replication. Proc. nat. Acad. Sci. (Wash.) **68**, 2806—2809 (1971).
Philipson, L., U. Lindberg, T. Persson, and B. Vennström: Transcription and processing of adenovirus RNA in productive infection. In: Advanc. in the Biosciences, Vol. 11 (G. Raspé, ed.), pp. 167—183. Pergamon Press, Vieweg, 1973.
Philipson, L., U. Pettersson, U. Lindberg, C. Tibbetts, B. Vennström, and T. Persson: RNA synthesis and processing in adenovirus infected cells. Cold Spr. Harb. Symp. quant. Biol. **39**, 447—456 (1974).
Piercy, S. E., and R. F. Sellers: Antibody response to a combined living attenuated distemper/hepatitis vaccine. Res. vet. Sci. **1**, 84—93 (1960).
Piña, M., and M. Green: Biochemical studies on adenovirus multiplication. IX. Chemical and base composition analysis of 28 human adenoviruses. Proc. nat. Acad. Sci. (Wash.) **54**, 547—551 (1965).
Piña, M., and M. Green: Base composition of the DNA of oncogenic simian adenovirus SA 7 and homology with human adenovirus DNAs. Virology **36**, 321—323 (1968).
Piña, M., and M. Green: Biochemical studies on adenovirus multiplication. XIV. Macromolecule and enzyme synthesis in cells replicating oncogenic and nononcogenic human adenoviruses. Virology **38**, 573—586 (1969).
Polasa, H., and M. Green: Biochemical studies on adenovirus multiplication. VIII. Analysis of protein synthesis. Virology **25**, 68—79 (1965).
Pontén, J.: Spontaneous and virus-induced transformation in cell culture. Virology Monographs, No. 8. Wien-New York: Springer, 1971.
Pope, J. H., and W. P. Rowe: Immunofluorescent studies of adenovirus 12 tumors and of cells transformed or infected by adenoviruses. J. exp. Med. **120**, 577—588 (1964).
Prage, L., and U. Pettersson: Structural proteins of adenoviruses. VII. Purification and properties of an arginine-rich core protein from adenovirus type 2 and type 3. Virology **45**, 364—373 (1971).
Prage, L., U. Pettersson, and L. Philipson: Internal basic proteins in adenovirus. Virology **36**, 508—511 (1968).
Prage, L., S. Höglund, and L. Philipson: Structural proteins of adenoviruses. VIII. Characterization of incomplete particles of adenovirus type 3. Virology **49**, 745 to 757 (1972).
Prage, L., E. Everitt, and L. Philipson: unpublished (1974).
Prage, L., U. Pettersson, S. Höglund, K. Lonberg-Holm, and L. Philipson: Structural proteins of adenoviruses. IV. Sequential degradation of the adenovirus type 2 virion. Virology **42**, 341—358 (1970).

PRICE, R., and S. PENMAN: Transcription of the adenovirus genome by an α-aminitine-sensitive ribonucleic acid polymerase in HeLa cells. J. Virol. **9**, 621—626 (1972 a).

PRICE, R., and S. PENMAN: A distinct RNA polymerase activity, synthesizing 5.5S, 5S and 4S RNA in nuclei from adenovirus 2-infected HeLa cells. J. molec. Biol. **70**, 435—450 (1972 b).

RABSON, A. S., R. L. KIRSCHSTEIN, and F. J. PAUL: Tumors produced by adenovirus 12 in mastomys and mice. J. nat. Cancer Inst. **32**, 77—87 (1964 a).

RABSON, A. S., G. T. O'CONOR, I. K. BEREZESKY, and F. J. PAUL: Enhancement of adenovirus growth in African green monkey kidney cell cultures by SV40. Proc. Soc. exp. Biol. (N.Y.) **116**, 187—190 (1964 b).

RAFAJKO, R. R.: Production and Standardization of adenovirus types 1 to 18 reference antisera. Amer. J. Hyg. **79**, 310—319 (1964).

RAFAJKO, R. R.: Routine establishment of serial lines of hamster embryo cells transformed by adenovirus type 12. J. nat. Cancer. Inst. **38**, 581—591 (1967).

RAPOZA, N. P.: A classification of simian adenoviruses based on hemagglutination. Amer. J. Epidem. **86**, 736—745 (1967).

RAPP, F.: The paraadenoviruses. Progr. exp. Tumor Res. (Basel) **18**, 104—137 (1973).

RAPP, F., J. S. BUTEL, and J. L. MELNICK: SV40-adenovirus "hybrid" populations: Transfer of SV40 determinants from one type of adenovirus to another. Proc. nat. Acad. Sci. (Wash.) **54**, 717—724 (1965).

RAPP, F., J. L. MELNICK, J. S. BUTEL, and T. KITÁHARA: The incorporation of SV40 genetic material into adenovirus 7 as measured by intranuclear synthesis of SV40 tumor antigen. Proc. nat. Acad. Sci. (Wash.) **52**, 1348—1352 (1964).

RAPP, F., M. JERKOFSKY, J. L. MELNICK, and B. LEVY: Variation in the oncogenic potential of human adenoviruses carrying a defective SV40 genome (PARA). J. exp. Med. **127**, 77—90 (1968).

RASKA, K., and W. A. STROHL: The response of BHK 21 cells to infection with type 12 adenovirus. VI. Synthesis of virus specific RNA. Virology **47**, 734—742 (1972).

RASKA, K., D. FROHWIRTH, and R. W. SCHLESINGER: Transfer ribonucleic acid in KB cells infected with adenovirus type 2. J. Virol. **5**, 464—469 (1970).

RASKA, K., L. PRAGE, and R. W. SCHLESINGER: The effects of arginine starvation on macromolecular synthesis in infection with type 2 adenovirus. II. Synthesis of virus-specific RNA and DNA. Virology **48**, 472—484 (1972).

RASKAS, H. J., and C. K. OKUBO: Transcription of viral RNA in KB cells infected with adenovirus type 2. J. Cell Biol. **49**, 438—449 (1971).

RASKAS, H. J., and S. BHADURI: Poly (adenylic acid) sequences in adenovirus ribonucleic acid released from isolated nuclei. Biochemistry **12**, 920—925 (1973).

RASKAS, J. H., D. C. THOMAS, and M. GREEN: Biochemical studies on adenovirus multiplication. XVII. Ribosome synthesis in uninfected and infected KB cells. Virology **40**, 893—902 (1970).

REEVES, W. C., R. P. SCRIVANI, W. E. PUGH, and W. P. ROWE: Recovery of an adenovirus from a feral rodent *Peromyscus maniculatus*. Proc. Soc. exp. Biol. (N.Y.) **124**, 1173—1175 (1967).

REICH, P. R., S. G. BAUM, J. A. ROSE, W. P. ROWE, and S. M. WEISSMAN: Nucleic acid homology studies of adenovirus type 7-SV40 interactions. Proc. nat. Acad. Sci. (Wash.) **55**, 336—341 (1966).

RIGGS, J. L., and E. H. LENNETTE: *In vitro* transformation of newborn-hamster kidney cells by simian adenoviruses. Proc. Soc. exp. Biol. (N.Y.) **126**, 802—806 (1967).

ROBINSON, A. J., H. B. YOUNGHUSBAND, and A. J. D. BELLETT: A circular DNA-protein complex from adenoviruses. Virology **56**, 54—69 (1973).

RODEN, A. T., H. G. PEREIRA, and D. M. CHAPRONIERE: Infection of volunteers by a virus (A.P.C. type 1) isolated from human adenoid tissue. Lancet **ii**, 592—596 (1956).

ROEDER, R. G., and W. J. RUTTER: Specific nucleolar and nucleoplasmic RNA polymerases. Proc. nat. Acad. Sci. (Wash.) **65**, 675—682 (1970).

ROSEN, L.: Hemagglutination of adenoviruses. Virology **5**, 574—577 (1958).

ROSEN, L.: Hemagglutination-inhibition technique for typing adenovirus. Amer. J. Hyg. **71**, 120—128 (1960).

Rosen, L., J. F. Hovis, and J. A. Bell: Further observations on typing adenovirus and a description of two possible additional serotypes. Proc. Soc. exp. Biol. (N.Y.) **110**, 710—713 (1962).

Rouse, H. C., and R. W. Schlesinger: An arginine-dependent step in the maturation of type 2 adenovirus. Virology **33**, 513—522 (1967).

Rouse, H. C., and R. W. Schlesinger: The effects of arginine starvation on macromolecular synthesis in infection with type 2 adenovirus. I. Synthesis and utilization of structural proteins. Virology **48**, 463—471 (1972).

Rowe, W. P.: Studies of adenovirus-SV 40 hybrid viruses. III. Transfer of SV 40 gene between adenovirus types. Proc. nat. Acad. Sci. (Wash.) **54**, 711—716 (1965).

Rowe, W. P., and S. G. Baum: Evidence for a possible genetic hybrid between adenovirus type 7 and SV 40 viruses. Proc. nat. Acad. Sci. (Wash.) **52**, 1340—1347 (1964).

Rowe, W. P., and S. G. Baum: Studies of adenovirus-SV 40 hybrid viruses. II. Defectiveness of the hybrid particles. J. exp. Med. **122**, 955—966 (1965).

Rowe, W. P., and W. E. Pugh: Studies of an adenovirus-SV 40 hybrid virus. V. Evidence for linkage between adenovirus and SV 40 genetic materials. Proc. nat. Acad. Sci. (Wash.) **55**, 1126—1132 (1966).

Rowe, W. P., and A. M. Lewis, Jr.: Serologic surveys for viral antibodies in cancer patients. Cancer Res. **28**, 19—20 (1968).

Rowe, W. P., J. W. Hartley, B. Roizman, and H. B. Levey: Characterization of a factor formed in the course of adenovirus infection of tissue cultures causing detachment of cells from glass. J. exp. Med. **108**, 713—729 (1958).

Rowe, W. P., R. J. Huebner, L. K. Gillmore, R. H. Parrott, and T. G. Ward: Isolation of a cytopathogenic agent from human adenoids undergoing spontaneous degeneration in tissue culture. Proc. Soc. exp. Biol. (N.Y.) **84**, 570—573 (1953).

Rowe, W. P., R. J. Huebner, J. W. Hartley, T. G. Ward, and R. H. Parrott: Studies of the adenoidal-pharyngeal-conjunctival (APC) group of viruses. Amer. J. Hyg. **61**, 197—218 (1955).

Russell, W. C., and B. Knight: Evidence for a new antigen within the adenovirus capsid. J. gen. Virol. **1**, 523—528 (1967).

Russell, W. C., and Y. Becker: A maturation factor for adenovirus. Virology **35**, 18—27 (1968).

Russell, W. C., and J. J. Skehel: The polypeptides of adenovirus-infected cells. J. gen. Virol. **15**, 45—47 (1972).

Russell, W. C., K. Hayashi, P. J. Sanderson, and H. G. Pereira: Adenovirus antigens — A study of their properties and sequential development in infection. J. gen. Virol. **1**, 495—507 (1967 a).

Russell, W. C., R. C. Valentine, and H. G. Pereira: The effect of heat on the anatomy of the adenovirus. J. gen. Virol. **1**, 509—522 (1967 b).

Russell, W. C., W. G. Laver, and P. J. Sanderson: Internal components of adenovirus. Nature (Lond.) **219**, 1127—1130 (1968).

Russell, W. C., C. Newman, and J. F. Williams: Characterization of temperature-sensitive mutants of adenovirus type 5 — serology. J. gen. Virol. **17**, 265—279 (1972 a).

Russell, W. C., J. J. Skehel, R. Machado, and H. G. Pereira: Phosphorylated polypeptides in adenovirus-infected cells. Virology **50**, 931—934 (1972 b).

Salk, J. E., U. Krech, J. S. Youngner, B. L. Bennett, L. J. Lewis, and P. L. Bazeley: Formaldehyde treatment and safety testing of experimental poliomyelitis vaccines. Amer. J. publ. Hlth **44**, 563—570 (1954).

Salzberg, S., and H. J. Raskas: Surface changes of human cells productively infected with human adenoviruses. Virology **48**, 631—637 (1972).

Samarina, O. P., E. M. Lukanidin, J. Molnar, and G. P. Georgiev: Structural organization of nuclear complexes containing DNA-like RNA. J. molec. Biol. **33**, 251—263 (1968).

Sambrook, J., B. Sugden, W. Keller, and P. A. Sharp: Transcription of simian virus 40. III. Orientation of RNA synthesis and mapping of early and late species of viral RNA extracted from lytically infected cells. Proc. nat. Acad. Sci. (Wash.) **70**, 3711—3715 (1973).

SAMBROOK, J., P. A. SHARP, B. OZANNE, and U. PETTERSSON: Studies on the transcription of simian virus 40 and adenovirus type 2. In: Control of Transcription (B. BISWAS, R. MANDAL, A. STEVENS, and W. COHN, eds.), pp. 167—180. Plenum Press, 1974.

SARMA, P. S., R. J. HUEBNER, and W. T. LANE: Induction of tumors in hamsters with an avian adenovirus (CELO). Science **149**, 1108 (1965).

SARMA, P. S., W. VASS, R. J. HUEBNER, H. IGEL, W. T. LANE, and H. C. TURNER: Induction of tumors in hamsters with infectious canine hepatitis virus. Nature (Lond.) **215**, 293—294 (1967).

SCHELL, K., and M. SCHMIDT: Adenovirus transformation of hamster embryo cells. I. Assay conditions. Arch. ges. Virusforsch. **24**, 332—341 (1968).

SCHELL, K., J. MARYAK, and M. SCHMIDT: Adenovirus transformation of hamster embryo cells. III. Maintenance conditions. Arch. ges. Virusforsch. **24**, 352—360 (1968 a).

SCHELL, K., J. MARYAK, J. YOUNG, and M. SCHMIDT: Adenovirus transformation of hamster embryo cells. II. Inoculation conditions. Arch. ges. Virusforsch. **24**, 342 to 351 (1968 b).

SCHELL, K., W. T. LANE, M. J. CASEY, and R. J. HUEBNER: Potentiation of oncogenicity of adenovirus type 12 grown in African green monkey kidney cell cultures preinfected with SV 40 virus: Persistence of both T antigens in the tumors and evidence for possible hybridization. Proc. nat. Acad. Sci. (Wash.) **55**, 81—88 (1966).

SCHLESINGER, R. W.: Adenoviruses: The nature of the virion and of controlling factors in productive or arbortive infection and tumorigenesis. Advanc. Virus Res. **14**, 1—61 (1969).

SCHMIDT, N. J., C. J. KING, and E. H. LENNETTE: Hemagglutination and hemagglutination-inhibition with adenovirus type 12. Proc. Soc. exp. Biol. (N.Y.) **118**, 208—211 (1965).

SCHOCHETMAN, G., and R. P. PERRY: Characterization of the messenger RNA released from L cell polyribosomes as a result of temperature shock. J. molec. Biol. **63**, 577—590 (1972).

SCHREIER, M. H., and T. STAEHELIN: Initiation of mammalian protein synthesis: The importance of ribosome and initiation factor quality for the efficiency of *in vitro* systems. J. molec. Biol. **73**, 329—349 (1973).

SELIVANOV, A. A., R. A. PLESHANOVA, E. A. SKRYABINA, and A. A. SMORODINTSEV: Testing the effectiveness of live adenovirus vaccine. I. Reactogenic properties. Acta virol. **8**, 263—270 (1964).

SHARP, P. A., B. SUGDEN, and J. SAMBROOK: Detection of two restriction endonuclease activities in *H. parainfluenzae* using analytical agarose-ethidium bromide electrophoresis. Biochemistry **12**, 3055—3063 (1973).

SHARP, P. A., P. H. GALLIMORE, and S. J. FLINT: Titration of viral RNA sequences in adenovirus 2 lytically infected cells and transformed cell lines. Cold Spr. Harb. Symp. quant. Biol. **39**, 457—474 (1974 a).

SHARP, P. A., U. PETTERSSON, and J. SAMBROOK: Viral DNA in transformed cells. I. A study of the sequences of adenovirus DNA in a line of transformed rat cells using specific fragments of the viral genome. J. molec. Biol. **86**, 709—726 (1974 b).

SHARPE, H. B.: Intranuclear inclusion bodies in the intestinal epithelium of pig foetuses experimentally infected with porcine adenovirus. J. Path. Bact. **93**, 353 to 355 (1967).

SHARPE, H. B. A., and D. M. JESSETT: Experimental infection of pigs with 2 strains of porcine adenovirus. J. comp. Path. **77**, 45—50 (1967).

SHEVLIAGHYN, V. J., and N. V. KARAZAS: Transformation of human cells by polyoma and Rous sarcoma viruses mediated by inactivated Sendai virus. Int. J. Cancer **6**, 234—244 (1970).

SHIMOJO, H., H. YAMAMOTO, and C. ABE: Differentiation of adenovirus 12 antigens in cultured cells. Virology **31**, 748—752 (1967).

Shiroki, K., J. Irisawa, and H. Shimojo: Isolation and preliminary characterization of temperature-sensitive mutants of adenovirus 12. Virology **49**, 1—11 (1972).

Shortridge, K. F., and F. Biddle: The proteins of adenovirus type 5. Arch. ges. Virusforsch. **29**, 1—24 (1970).

Singer, R. H., and S. Penman: Stability of HeLa cell mRNA in actinomycin. Nature (Lond.) **240**, 100—102 (1972).

Sinha, S. K., L. W. Fleming, and S. Scholes: Current considerations in public health of the role of animals in relation to human viral diseases. J. Amer. vet. med. Ass. **136**, 481—485 (1960).

Sjögren, H. O., J. Minowada, and J. Ankerst: Specific transplantation antigens of mouse sarcoma induced by adenovirus type 12. J. exp. Med. **125**, 689—701 (1967).

Slifkin, M., L. Merkow, and N. P. Rapoza: Tumor induction by simian adenovirus 30 and establishment of tumor cell lines. Cancer Res. **28**, 1173—1179 (1968).

Smith, K. O., and J. L. Melnick: Adenovirus-like particles from cancers induced by adenovirus-12 but free of infectious virus. Science **145**, 1190—1192 (1964).

Smith, K. O., W. D. Gehle, and M. D. Trousdale: Architecture of the adenovirus capsid. J. Bact. **90**, 254—261 (1965).

Smith, K. O., W. D. Gehle, and W. T. Kniker: Serologic evidence that infectious canine hepatitis virus commonly infects humans and is related to human adenovirus type 8. J. Immunol. **105**, 1036—1039 (1970).

Soeiro, R., M. H. Vaughan, J. R. Warner, and J. E. Darnell: The turnover of nuclear DNA-like RNA in HeLa cells. J. Cell Biol. **39**, 112—118 (1968).

Sohier, R., Y. Chardonnet et M. Purnieras: Isolement d'un adénovirus type 1 à partir d'une adénopathie cervicale maligne. Presse méd. **71**, 1733—1734 (1963).

Sohier, R., Y. Chardonnet, and M. Prunieras: Adenoviruses: Status of current knowledge. In: Progr. med. Virol. (J. Melnick, ed.), Vol. 7, 253—325 (1965).

Spirin, A. S., and M. Nemer: Messenger RNA in early sea-urchin embryos: cytoplasmic particles. Science **150**, 214—217 (1965).

Stasny, J. T., A. R. Neurath, and B. A. Rubin: Effect of formamide on the capsid morphology of adenovirus types 4 and 7. J. Virol. **2**, 1429—1442 (1968).

Stevens, D. A., M. Schaeffer, J. P. Fox, C. D. Brandt, and M. Romano: Standardization and certification of reference antigens and antisera for 30 human adenovirus serotypes. Amer. J. Epidem. **86**, 617—633 (1967).

Strohl, W. A.: The response of BHK 21 cells to infection with type 12 adenovirus. I. Cell killing and T antigen synthesis as correlated viral genome functions. Virology **39**, 642—652 (1969 a).

Strohl, W. A.: The response of BHK 21 cells to infection with type 12 adenovirus. II. Relationship of virus-stimulated DNA synthesis to other viral functions. Virology **39**, 653—665 (1969 b).

Strohl, W. A., H. C. Rouse, and R. W. Schlesinger: Properties of cells derived from adenovirus-induced hamster tumors by long-term *in vitro* cultivation. II. Nature of the restricted response to type 2 adenovirus. Virology **28**, 645—658 (1966).

Strohl, W. A., A. S. Rabson, and H. Rouse: Adenovirus tumorigenesis: Role of the viral genome in determining tumor morphology. Science **156**, 1631—1633 (1967).

Strohl, W. A., H. Rouse, K. Teets, and R. W. Schlesinger: The response of BHK 21 cells to infection with type 12 adenovirus. III. Transformation and restricted replication of superinfecting type 2 adenovirus. Arch. ges. Virusforsch. **31**, 93—112 (1970).

Studdert, M. J., C. R. Wilks, and L. Coggins: Antigenic comparisons and serologic survey of equine adenoviruses. Amer. J. vet. Res. **35**, 693—699 (1974).

Sugino, A., S. Hirose, and R. Okazaki: RNA-linked nascent DNA fragments in *Escherichia coli*. Proc. nat. Acad. Sci. (Wash.) **69**, 1863—1867 (1972).

Summers, D. F., and J. V. Maizel: Evidence for large precursor proteins in poliovirus synthesis. Proc. nat. Acad. Sci. (Wash.) **59**, 966—971 (1968).

Sundquist, B., U. Pettersson, L. Thelander, and L. Philipson: Structural proteins of adenovirus. IX. Molecular weight and subunit composition of the adenovirus type 2 fiber. Virology **51**, 252—256 (1973 a).

SUNDQUIST, B., E. EVERITT, L. PHILIPSON, and S. HÖGLUND: Assembly of adenoviruses. J. Virol. **11**, 449—459 (1973 b).

SUSSENBACH, J. S.: Early events in the infection process of adenovirus type 5 in HeLa cells. Virology **33**, 567—574 (1967).

SUSSENBACH, J.S.: On the fate of adenovirus DNA in KB cells. Virology **46**, 969—972 (1971).

SUSSENBACH, J. S., and P. VAN DER VLIET: Characterization of adenovirus DNA in cells infected with adenovirus type 12. Virology **48**, 224—229 (1972).

SUSSENBACH, J. S., and P. C. VAN DER VLIET: Studies on the mechanism of replication of adenovirus DNA. I. The effect of hydroxyurea. Virology **54**, 299—303 (1973).

SUSSENBACH, J. S., D. J. ELLENS, and H. S. JANSZ: Studies on the mechanism of replication of adenovirus DNA. II. The nature of single-stranded DNA in replicative intermediates. J. Virol. **12**, 1131—1138 (1973).

SUSSENBACH, J. S., P. C. VAN DER VLIET, D. J. ELLENS, and H. S. JANSZ: Linear intermediates in the replication of adenovirus DNA. Nature (New Biol.) **239**, 47—49 (1972).

SUZUKI, E., and H. SHIMOJO: A temperature-sensitive mutant of adenovirus 31, defective in viral deoxyribonucleic acid. Virology **43**, 488—494 (1971).

SWANGO, L. J., G. A. EDDY, and L. N. BINN: Serologic comparisons of infectious canine hepatitis and Toronto A 26/61 canine adenoviruses. Amer. J. vet. Res. **30**, 1381—1387 (1969).

SZYBALSKI, W., K. HUBINSKI, Z. HRADECNA, and W. C. SUMMERS: Analytical and preparative separation of the complementary DNA strands. In: Methods in Enzymology, Vol. XXI (L. GROSSMAN and K. MOLDAVE, eds.), pp. 383—413. New York: Academic Press, 1971.

TAKAHASHI, M., S. VEDA, and T. OGINO: Enhancement of thymidine kinase activity of human embryonic kidney cells and newborn hamster kidney cells by infection with human adenovirus types 5 and 12. Virology **30**, 742—743 (1966).

TAKEMORI, N.: Genetic studies with tumorigenic adenoviruses. III. Recombination in adenovirus type 12. Virology **47**, 157—167 (1972).

TAKEMORI, N., J. L. RIGGS, and C. ALDRICH: Genetic studies with tumorigenic adenoviruses. I. Isolation of cytocidal (cyt)mutants of adenovirus type 12. Virology **36**, 575—586 (1968).

TAVITIAN, A., S. C. URETSKY, and G. ACS: Selective inhibition of ribosomal RNA synthesis in mammalian cells. Biochim. biophys. Acta (Amst.) **157**, 33—42 (1968).

TAVITIAN, A., J. PERIES, J. CHUAT, and M. BOIRON: Estimation of the molecular weight of adenovirus 12 tumor CF antigen by rate-zonal centrifugation. Virology **31**, 719 to 721 (1967).

TEJG-JENSEN, B., B. FURUGREN, I. LINDQVIST und L. PHILIPSON: Röntgenkleinwinkelstreuung an Hexon aus Adenovirus Typ 2. Mh. Chem. **103**, 1730—1736 (1972).

THOMAS, D. C., and M. GREEN: Biochemical studies on adenovirus multiplication. XI. Evidence of a cytoplasmic site for the synthesis of viral-coded proteins. Proc. nat. Acad. Sci. (Wash.) **56**, 243—246 (1966).

TIBBETTS, C., and U. PETTERSSON: Complementary strand-specific sequences from unique fragments of adenovirus type 2 DNA for hybridization-mapping experiments. J. molec. Biol. **88**, 767—784 (1974).

TIBBETTS, C., K. JOHANSSON, and L. PHILIPSON: Hydroxylapatite chromatography and formamide denaturation of adenovirus DNA. J. Virol. **12**, 218—225 (1973).

TIBBETTS, C., U. PETTERSSON, K. JOHANSSON, and L. PHILIPSON: Relationship of mRNA from productively infected cells to the complementary strands of adenovirus type 2 DNA. J. Virol. **13**, 370—377 (1974).

TIMONEY, P. J.: Adenovirus precipitating antibodies in the sera of some domestic animal species in Ireland. Brit. vet. J. **127**, 567—571 (1971).

TOCKSTEIN, G., H. POLASA, M. PIÑA, and M. GREEN: A simple purification procedure for adenovirus type 12 T and tumor antigens and some of their properties. Virology **36**, 377—386 (1968).

TODARO, G. J., and S. A. AARONSON: Human cell strains susceptible to focus formation by human adenovirus type 12. Proc. nat. Acad. Sci. (Wash.) **61**, 1272—1278 (1968).

TODD, J. D.: Comments on rhinoviruses and parainfluenza viruses of horses. J. Amer. vet. med. Ass. **155**, 387—380 (1969).

TONEGAWA, S., G. WALTER, A. BERNARDINI, and R. DULBECCO: Transcription of the SV 40 genome in transformed cells and during lytic infection. Cold Spr. Harb. Symp. quant. Biol. **35**, 823—832 (1970).

TOOZE, J. (ed.): The adenoviruses. In: Molecular Biology of Tumor Viruses, pp. 420—469. Cold Spring Harbor Laboratory Press, 1973.

TRENTIN, J. J., and E. BRYAN: Immunization of hamsters and histogenic mice against transplantation of tumors induced by human adenovirus type 12. Proc. Amer. Ass. Cancer Res. **5**, 64 (1964).

TRENTIN, J. J., Y. YABE, and G. TAYLOR: The quest for human cancer viruses. Science **137**, 835—841 (1962).

TRENTIN, J. J., G. L. VAN HOOSIER, JR., and L. SAMPER: The oncogenicity of human adenoviruses in hamsters. Proc. Soc. exp. Biol. (N.Y.) **127**, 683—689 (1968).

TSEUI, D., K. FUJINAGA, and M. GREEN: The mechanism of viral carcinogenesis by DNA mammalian viruses: RNA transcripts containing viral and highly reiterated cellular base sequences in adenovirus-transformed cells. Proc. nat. Acad. Sci. (Wash.) **69**, 427—430 (1972).

USTACELEBI, S., and J. T. WILLIAMS: Temperature-sensitive mutants of adenovirus defective in interferon induction at non-permissive temperature. Nature (Lond.) **235**, 52—53 (1972).

VALENTINE, R. C., and H. G. PEREIRA: Antigens and structure of the adenovirus. J. molec. Biol. **13**, 13—20 (1965).

VAN DER EB, A. J.: Intermediates in type 5 adenovirus DNA replication. Virology **51**, 11—23 (1973).

VAN DER EB, A. J., L. W. KESTERN, and E. F. J. VAN BRUGGEN: Structural properties of adenovirus DNA's. Biochim. biophys. Acta (Amst.) **182**, 530—541 (1969).

VAN DER NORDAA, J.: Transformation of rat kidney cells by adenovirus type 12. J. gen. Virol. **2**, 269—272 (1968 a).

VAN DER NORDAA, J.: Transformation of rat cells by adenovirus types 1, 2 and 3. J. gen. Virol. **3**, 303—304 (1968 b).

VAN DER VEEN, J., and A. MES: Serological classification of two mouse adenoviruses- Arch. ges. Virusforsch. **45**, 386—387 (1974).

VAN DER VLIET, P. C., and J. S. SUSSENBACH: The mechanism of adenovirus-DNA-synthesis in isolated nuclei. Europ. J. Biochem. **30**, 548—592 (1972).

VAN DER VLIET, P. C., and A. J. LEVINE: DNA-binding proteins specific for cells infected by adenovirus. Nature (New Biol.) **246**, 170—174 (1973).

VELICER, L., and H. GINSBERG: Cytoplasmic synthesis of type 5 adenovirus capsid proteins. Proc. nat. Acad. Sci. (Wash.) **61**, 1264—1271 (1968).

VELICER, L., and H. GINSBERG: Synthesis, transport and morphogenesis of type 5 adenovirus capsid proteins. J. Virol. **5**, 338—352 (1970).

WADELL, G.: Hemagglutination with adenovirus serotypes belonging to Rosen's subgroup II and III. Proc. Soc. exp. Biol. (N.Y.) **132**, 413—421 (1969).

WADELL, G.: Structural and biological properties of capsid components of human adenoviruses. Ph. D. thesis, Karolinska Institute, Stockholm, 1970.

WADELL, G.: Sensitization and neutralization of adenovirus by specific sera against capsid subunits. J. Immunol. **108**, 622—632 (1972).

WADELL, G., and E. NORRBY: The soluble hemagglutinins of adenoviruses belonging to Rosen's subgroup III. I. The rapidly sedimenting hemagglutinin. Arch. ges. Virusforsch. **26**, 53—62 (1969 a).

WADELL, G., and E. NORRBY: Immunological and other biological characteristics of pentons of human adenoviruses. J. Virol. **4**, 671—680 (1969 b).

WADELL, G., E. NORRBY, and P. SKAARET: The soluble hemagglutinins of adenoviruses belonging to Rosen's subgroup III. I. The rapidly sedimenting hemagglutinin. Arch. ges. Virusforsch. **26**, 32—52 (1969).

WADELL, G., M.-L. HAMMARSKJÖLD, and T. VARSANYI: Incomplete virus particles of adenovirus type 16. J. gen. Virol. **20**, 287—303 (1973).

Wall, R., L. Philipson, and J. E. Darnell: Processing of adenovirus specific nuclear RNA during virus replication. Virology 50, 27—34 (1972).

Wall, R., J. Weber, Z. Gage, and J. E. Darnell: Production of viral mRNA in adenovirus-transformed cells by the post-transcriptional processing of heterogeneous nuclear RNA containing viral and cell sequences. J. Virol. 11, 953—960 (1973).

Wallace, R. D., and J. Kates: On the state of the adenovirus 2 DNA in the nucleus and its mode of transcription. Studies with viral protein complexes and isolated nuclei. J. Virol. 9, 627—635 (1972).

Walter, G., and J. V. Maizel, Jr.: The polypeptides of adenovirus. IV. Detection of early and late virus-induced polypeptides and their distribution in subcellular fractions. Virology 57, 402—408 (1974).

Warocquier, R., J. Samaille, and M. Green: Biochemical studies on adenovirus multiplication. XVI. Transcription of the adenovirus genome during abortive infection at elevated temperatures. J. Virol. 4, 423—428 (1969).

Wasmuth, E. H., and A. A. Tytell: Physical studies with adenovirus hexon antigens. Life Sci. 6, 1063—1068 (1966).

Watson, J. D.: Origin of concatemeric T7 DNA. Nature (New Biol.) 239, 197—201 (1972).

Watson, J. D., and F. H. C. Crick: The structure of DNA, Cold Spr. Harb. Symp. quant. Biol. 18, 123—131 (1953).

Weber, J.: Absence of adenovirus-specific repressor in adenovirus tumour cells. J. gen. Virol. 22, 259—264 (1974).

Westphal, H., and R. Dulbecco: Viral DNA in polyoma and SV 40-transformed cell lines. Proc. nat. Acad. Sci. (Wash.) 59, 1158—1165 (1968).

Whitcutt, J. M., and H. S. Gear: Transformation of newborn hamster cells with simian adenovirus SA 7. Int. J. Cancer 3, 566—571 (1968).

White, D. O., M. D. Scharff, and J. V. Maizel, Jr.: The polypeptides of adenoviruses. III. Synthesis in infected cells. Virology 38, 395—406 (1969).

Wickner, W., D. Brutlag, R. Schekman, and A. Kornberg: RNA synthesis initiates in vitro conversion of M 13 DNA to its replicative form. Proc. nat. Acad. Sci. (Wash.) 69, 965—969 (1972).

Wigand, R.: Adenovirus types 20, 25 and 28: atypical members of group II. J. gen. Virol. 6, 325—328 (1969).

Wigand, R., H. Bauer, F. Lang, and W. Adam: Neutralization of the adenoviruses types 1 to 28: Specificity and antigenic relationships. Arch. ges. Virusforsch. 15, 188—199 (1965).

Wilcox, G. E.: Isolation of adenoviruses from cattle with conjunctivitis and keratoconjunctivitis. Aust. vet. J. 45, 265—270 (1969).

Wilcox, W. C., and H. S. Ginsberg: Purification and immunological characterization of types 4 and 5 adenovirus soluble antigens. Proc. nat. Acad. Sci. (Wash.) 47, 512 to 526 (1961).

Wilcox, W. C., and H. S. Ginsberg: Structure of type 5 adenovirus. I. Antigenic relationship of virus structural proteins to virus specific soluble antigens from infected cells. J. exp. Med. 118, 295—306 (1963 a).

Wilcox, W. C., and H. S. Ginsberg: Production of specific neutralizing antibody with soluble antigens of type 5 adenovirus. Proc. Soc. exp. Biol. (N.Y.) 114, 37—42 (1963 b).

Wilcox, W. C., and H. S. Ginsberg: Protein synthesis in type 5 adenovirus infected cells. Effect of p-fluorophenylalanine on synthesis of protein nucleic acids and infectious virus. Virology 20, 269—280 (1963 c).

Wilcox, W. C., H. S. Ginsberg, and T. F. Anderson: Structure of type 5 adenovirus. II. Fine structure of virus subunits. Morphological relationship of structural subunits to virus-specific soluble antigens from infected cells. J. exp. Med. 118, 307 314 (1963).

Wilhelm, J. M., and H. S. Ginsberg: Synthesis in vitro of type 5 adenovirus capsid proteins. J. Virol. 9, 973—980 (1972).

WILKIE, N. M., S. USTACELEBI, and J. F. WILLIAMS: Characterization of temperature-sensitive mutants of adenovirus type 5. Nucleic acid synthesis. Virology **51**, 499—503 (1973).

WILLEMS, M., M. PENMAN, and S. PENMAN: The regulation of RNA synthesis and processing in the nucleolus during inhibition of protein synthesis. J. Cell Biol. **41**, 177 to 187 (1969).

WILLIAMS, J. F.: Oncogenic transformation of hamster embryo cells *in vitro* by adenovirus type 5. Nature (Lond.) **243**, 162—163 (1973).

WILLIAMS, J. E., and S. USTACELEBI: Complementation and recombination with temperature-sensitive mutants of adenovirus type 5. J. gen. Virol. **13**, 345—348 (1971).

WILLIAMS, J. F., C. S. YOUNG, and P. E. AUSTIN: Genetic analysis of human adenovirus type 5 in permissive and non-permissive cells. Cold Spr. Harb. Symp. quant. Biol. **39**, 427—437 (1974).

WILLIAMS, J. F., M. GHARPURE, S. USTACELEBI, and S. McDONALD: Isolation of temperature-sensitive mutants of adenovirus type 5. J. gen. Virol. **11**, 95—102 (1971).

WILLS, E. J., W. C. RUSSELL, and J. F. WILLIAMS: Adenovirus-induced crystals: Studies with temperature-sensitive mutants. J. gen. Virol. **20**, 407—412 (1973).

WILNER, B. I.: A classification of the major groups of human and other animal viruses, pp. 120—132. Minneapolis. Minn.: Burgess, 1969.

WINNACKER, E.: Adenovirus type 2 DNA replication. I. Evidence for discontinuous DNA synthesis. J. Virol. in press (1975).

WINNACKER, E.-L., G. MAGNUSSON, and P. REICHARD: Replication of polyoma DNA in isolated nuclei. I. Characterization of the system from mouse fibroblast 3T 6 cells. J. molec. Biol. **72**, 523—537 (1972).

WINTERS, W. D., and W. C. RUSSELL: Studies on assembly of adenovirus *in vitro*. J. gen. Virol. **10**, 181—194 (1971).

WINTERS, W. D., A. BROWNSTONE, and H. G. PEREIRA: Separation of adenovirus penton base antigens by preparative gel electrophoresis. J. gen. Virol. **9**, 105—110 (1970).

WOLFSON, J., and D. DRESSLER: Adenovirus 2-DNA contains an inverted terminal repetition. Proc. nat. Acad. Sci. (Wash.) **69**, 3054—3057 (1972).

YABE, Y., L. SAMPER, G. TAYLOR, and J. J. TRENTIN: Cancer induction in hamsters by human type 12-adenovirus. Effect of route of injection. Proc. Soc. exp. Biol. (N.Y.) **113**, 221—224 (1963).

YABE, Y., L. SAMPER, E. BRYAN, G. TAYLOR, and J. J. TRENTIN: Oncogenic effect of human adenovirus type 12 in mice. Science **143**, 46—47 (1964).

YOHN, D. S.: Sex-related resistance in hamsters to adenovirus oncogenesis. Progr. exp. Tumor Res. (Basel) **18**, 138—165 (1973).

YOHN, D. S., C. A. FUNK, and J. T. GRACE, JR.: Sex-related resistance in hamsters to adenovirus-12 oncogenesis. II. Influence of virus dose. J. Virol. **1**, 1186—1192 (1967).

YOHN, D. S., L. WEISS, and M. E. NEIDERS: A comparison of the distribution of tumors produced by intravenous injection of type 12 adenovirus and adeno-12 tumor cells. Cancer Res. **28**, 571—576 (1968).

YOHN, D. S., C. A. FUNK, V. I. KALNINS, and J. T. GRACE, JR.: Sex-related resistance in hamsters to adenovirus-12 oncogenesis. Influence of thymectomy at three weeks of age. J. nat. Cancer. Inst. **35**, 717—624 (1965).

ZAIN, B. S., R. DHAR, S. M. WEISSMAN, P. LEBOWITZ, and A. M. LEWIS, JR.: Preferred site for initiation of RNA transcription by *Escherichia coli* RNA polymerase within the simian virus 40 DNA segment of the nondefective adenovirus-simian virus 40 hybrid viruses Ad2^+ND$_1$ and Ad2^+ND$_3$. J. Virol. **11**, 682—693 (1973).

ZIMMERMAN, J. E., and K. RASKA: Inhibition of adenovirus type 12 induced DNA synthesis in G 1-arrested BHK 21 cells by dibutyryl adenosine cyclic 3′, 5′-monophosphate. Nature (New Biol.) **239**, 145—147 (1972).

ZIMMERMAN, J., K. RASKA, and W. A. STROHL: The response of BHK 21 cells to infection with type 12 adenovirus. IV. Activation of DNA-synthesizing apparatus. Virology **42**, 1147—1150 (1970).

Zur Hausen, H.: Induction of specific chromosomal aberrations by adenovirus type 12 in human embryonic kidney cells. J. Virol. **1**, 1174—1185 (1967).

Zur Hausen, H.: Interactions of adenovirus type 12 with host cell chromosomes. Progr. exp. Tumor Res. (Basel) **18**, 240—259 (1973).

Zur Hausen, H., and F. Sokol: Fate of adenovirus type 12 genomes in nonpermissive cells. J. Virol. **4**, 256—271 (1969).

Zylber, E. A., and S. Penman: Products of RNA polymerase in HeLa cell nuclei. Proc. nat. Acad. Sci. (Wash.) **68**, 2861—2865 (1971).

VIROLOGY MONOGRAPHS

Volume 1:

ECHO Viruses
By **H. A. Wenner** and **A. M. Behbehani**

Reoviruses
By **L. Rosen**

4 figures. IV, 107 pages. 1968.

ISBN 3-211-80889-2 (Wien)
ISBN 0-387-80889-2 (New York)

Volume 2:

The Simian Viruses
By **R. N. Hull**

Rhinoviruses
By **D. A. J. Tyrrell**

19 figures. IV, 124 pages. 1968.

ISBN 3-211-80890-6 (Wien)
ISBN 0-387-80890-6 (New York)

Volume 3:

Cytomegaloviruses
By **J. B. Hanshaw**

Rinderpest Virus
By **W. Plowright**

Lumpy Skin Disease Virus
By **K. E. Weiss**

26 figures. IV, 131 pages. 1968.

ISBN 3-211-80891-4 (Wien)
ISBN 0-387-80891-4 (New York)

Volume 4:

The Influenza Viruses
By **L. Hoyle**

58 figures. IV, 375 pages. 1968.

ISBN 3-211-80892-2 (Wien)
ISBN 0-387-80892-2 (New York)

Volume 5:

Herpes Simplex and Pseudorabies Viruses
By **A. S. Kaplan**

14 figures. IV, 115 pages. 1969.

ISBN 3-211-80932-5 (Wien)
ISBN 0-387-80932-5 (New York)

Volume 6:

Interferon
By **J. Vilček**

4 figures. IV, 141 pages. 1969.

ISBN 3-211-80933-3 (Wien)
ISBN 0-387-80933-3 (New York)

Volume 7:

Polyoma Virus
By **B. E. Eddy**

Rubella Virus
By **E. Norrby**

22 figures. IV, 174 pages. 1969.

ISBN 3-211-80934-1 (Wien)
ISBN 0-387-80934-1 (New York)

VIROLOGY MONOGRAPHS

Volume 8:

Spontaneous and Virus Induced Transformation in Cell Culture

By **J. Pontén**

35 figures. IV, 253 pages. 1971.

ISBN 3-211-80991-0 (Wien)
ISBN 0-387-80991-0 (New York)

Volume 9:

African Swine Fever Virus

By **W. R. Hess**

Bluetongue Virus

By **P. G. Howell** and
D. W. Verwoerd

5 figures. IV, 74 pages. 1971.

ISBN 3-211-81006-4 (Wien)
ISBN 0-387-81006-4 (New York)

Volume 10:

Lymphocytic Choriomeningitis Virus

By **F. Lehmann-Grube**

16 figures. V, 173 pages. 1971.

ISBN 3-211-81017-X (Wien)
ISBN 0-387-81017-X (New York)

Volume 11:

Canine Distemper Virus

By **M. J. G. Appel** and
J. H. Gillespie

Marburg Virus

By **R. Siegert**

50 figures. IV, 153 pages. 1972.

ISBN 3-211-81059-5 (Wien)
ISBN 0-387-81059-5 (New York)

Volume 12:

Varicella Virus

By **D. Taylor-Robinson** and
A. E. Caunt

10 figures. IV, 88 pages. 1972.

ISBN 3-211-81065-X (Wien)
ISBN 0-387-81065-X (New York)

Volume 13:

Lactic Dehydrogenase Virus

By **K. E. K. Rowson** and
B. W. J. Mahy

54 figures. IV, 121 pages. 1975.

ISBN 3-211-81270-9 (Wien)
ISBN 0-387-81270-9 (New York)

Volume 14:

Molecular Biology of Adenoviruses

By **L. Philipson, U. Pettersson,** and **U. Lindberg**

20 figures. IV, 115 pages. 1975.

ISBN 3-211-81284-9 (Wien)
ISBN 0-387-81284-9 (New York)

QR
360
V52
no.14

MAY 4 1976